Manual of Best Management Practices (BMP`S) for Agriculture in New Hampshire

Best Management Practices for the Handling of Agricultural Compost, Fertilizer and Manure

United Book Prints

USDA

Reprinted in India by:

UNITED BOOK PRINTS
5A, Bhagat Ki Kothi,
Near Bhagat Ki Kothi Police Chowki
Jodhpur 342 001, India
E-mail: unitedbookprints@gmail.com

Print : 2021

ISBN: 978-93-83692-65-1
eISBN: 978-93-88148-43-6

Printed in India

MANUAL OF BEST MANAGEMENT PRACTICES (BMPS) FOR AGRICULTURE IN NEW HAMPSHIRE

Revised
June 2011

Nutrient Best Management Practices For Agricultural Nonpoint Source Pollution

Prepared by the
Agricultural Best Management Practices Task Force
and the
USDA Natural Resources Conservation Service (NRCS), Durham, NH
for the
New Hampshire Department of Agriculture, Markets, and Food
Concord, NH

FORWARD

This manual is a cooperative effort by several of New Hampshire's conservation partners. It presents guidance to landowners, town officials, state agencies, and others to help maintain the state's agricultural base and protect water quality. It discusses handling of manure, agricultural compost and chemical fertilizer. Handling is addressed in relation to farm operations, natural resource conservation, water quality, and human, animal and plant health. Nonpoint source pollution and resolution of agricultural environmental and social complaints are also discussed.

Agriculture is an important business in New Hampshire, producing food and fiber for local and regional use. Its importance is reflected in RSA 432:32-35 which limits nuisance liability of agricultural operations. It also provides secondary benefits to citizens and visitors alike. Open space, vistas, and recreation opportunities are available in greater numbers due to farming.

New Hampshire is also fortunate to have some of the best quality lakes and rivers in the United States. While New Hampshire's surface waters are important for recreation, both surface water and ground water are utilized for domestic water supplies. New Hampshire's economy, including farming, is dependent upon a healthy environment. Protecting this state's water resources is a major concern. The use of Best Management Practices for agriculture is an avenue to protect the quality of our lakes, streams, ground water and rivers for future generations.

Recognizing that the shorelands of the state are among its most valuable natural resources, and that the protection of these shorelands is essential to maintain the integrity of public waters, the New Hampshire General Court passed the Comprehensive Shoreland Protection Act (RSA 483-B) in 1991. Even though agricultural activities and operations are exempt from RSA 483-B, they must conform to best management practices determined by the USDA Natural Resources Conservation Service, the UNH Cooperative Extension and the New Hampshire Department of Agriculture Markets & Food. Persons engaging in these activities and operations in the protected shoreland shall work directly with the local representatives of the above agencies. The protected shoreland is all land within 250 feet of the public boundary line of public waters, as defined by the Act.

This Manual discusses and lists Best Management Practices for manure, agricultural compost and chemical fertilizer. As indicated in RSA 431:33-35, the practices for handling manure, agricultural compost and chemical fertilizer "...are based on the best available research and scientific data..." They are management, agronomic/vegetative and structural practices that permit economically viable production while achieving the least possible adverse impact upon the environment, including water quality. They also minimize possible adverse impacts on human, animal and plant health.

Lorraine S. Merrill
Commissioner
NH Department of Agriculture, Markets & Food

TABLE OF CONTENTS

INTRODUCTION

The land resource and farmers who use it represent the foundation of our Nation. Farming produces food, fiber, and other products for sustaining the state and country. In New Hampshire, about 3,400 commercial and many part-time and hobby farms produce food and fiber for local and regional markets. This farming can continue to be viable within the conservation needs of the farm, surrounding area and watershed. Continuous protection of the state's environmental quality can be assured by using Best Management Practices (BMPs). These agricultural BMPs are management, agronomic/vegetative and structural practices that permit economical and viable production while achieving the least possible adverse impact on the environment, including water quality. They also minimize possible adverse impacts on human, animal and plant health.

Best Management Practices prevent pollution from agricultural operations. Plant nutrients, bacteria, sediment and agricultural chemicals can be controlled so that pollution of surface and ground water does not occur and limit the use for drinking, aquatic life and recreation. Odor, vectors, and other nuisances can also be minimized by adequate BMPs.

This manual discusses water quality, nonpoint source pollution and the selection and use of BMPs for manure, agricultural compost, and chemical fertilizer. It provides lists of more common BMPs for preliminary consideration. The information sources in the reference section provide some guidance in selecting, planning, designing and implementing Best Management Practices. Professional judgment is required to properly select BMPs for a particular farm or site. It is not intended that all BMPs necessarily be applied to a particular situation. The manual also discusses agricultural water, air and nuisance complaints and positive steps to resolve them. Information is included on RSA 431:33-35, "Manure, Agricultural Compost, and Chemical Fertilizer Handling," and the complaint process.

AGRICULTURE AND THE ENVIRONMENT

Water Quality

The value of water lies in its usefulness for a wide variety of purposes, and the quality determines its acceptability for use. Quality is impacted when water is contaminated to a level where it is no longer acceptable for a particular use. Pollution, which limits the usefulness of receiving waters, has a significant effect on the environment. Therefore, maintaining or improving the quality of ground and surface water is important.

Potential ground water contaminants from agricultural operations include nutrients, generally nitrogen, agricultural chemicals, and bacteria. Potential surface water contaminants include agricultural chemicals attached to sediment, organic matter, bacteria, nutrients, including nitrogen, phosphorus, and sediment.

Under natural conditions, ground water tends to maintain a relatively constant quality over time. Soil filtration removes turbidity, color, and micro-organisms, depending on the soil and its chemical characteristics. Some chemicals are adsorbed depending on soil type. Because ground water is available throughout the state, it is often used for domestic supply. Ground water is also desirable because water recharging an aquifer has the potential to be purified naturally as it percolates through the soil. However, aquifers overlain by porous materials, such as sand and gravel, allow pollutants to move into the ground water.

In New Hampshire, surface water is often used for domestic purposes and there is concern for its quality. Surface water quality also has a considerable effect on recreational uses and on fish and other aquatic animals.

Water quality regulations relate to the physical and chemical properties of water as well as to toxic levels of natural and manufactured substances. Potable water has the most stringent quality requirements. Failure of domestic supplies to meet standards for even short periods of time can result in serious illness. Water quality standards also address aquatic life and recreation concerns.

Farms require a domestic water supply in addition to water used for a variety of other purposes. Livestock farmers are especially concerned with water quality for health and product quality. Farmers must be particularly careful that farm water supplies do not become contaminated.

Nonpoint Source Pollution

Improper or inadequate agricultural management activities can pollute surface and ground water resources. Potential agricultural pollution includes point and nonpoint source pollutants. Point source pollutants are discrete sources or where the specific point of entry of the pollutant is readily identified, such as a spill. Nonpoint source pollutants are diffuse in manner, with no definite point of entry and the source may not be readily discernible. In contrast to point sources, nonpoint sources generally result from precipitation, land runoff, or percolation. The impact to receiving waters is usually directly dependent on precipitation.

Nonpoint pollution sources are the most common for agriculture and are the ones generally discussed in this manual. Nonpoint pollution may be generated over a large area, such as a feedlot or field. These pollution sources cannot be easily treated with point-type treatment facilities. Therefore, agricultural nonpoint source pollution problems are solved by managing the potential sources and application sites.

Potential agricultural nonpoint source pollutants include nutrients, agricultural chemicals, organic wastes, and bacteria. Manure, agricultural compost, and chemical fertilizer are spread over field and crops as part of normal farming operations. Other potential pollutants such as petroleum may result from farming operations. Pollution can result if precipitation or runoff water detaches soil and other materials and transports them to surface water bodies or leaches them into ground water.

Several nonpoint sources may contribute to an accumulation of pollutants at the lower end of a watershed. Planning on a watershed scale may be necessary for these situations when determining practices necessary to solve water quality problems.

Other Impacts

Inadequate agricultural practices may have adverse impacts on air resources and social conditions in the community. For example, farming viewed by a neighbor who has a similar enterprise as compared to one who works in the city could be completely different. Appropriate practices will minimize social effects on the community. Emissions of ammonia and other gases from farming operations including livestock operations may degrade air quality if care is not taken. Odors, from confined livestock, waste storage areas, lagoons, and field application of wastes may be minimized to prevent offending the neighbors. Any hazards from disease and parasites, insects and other vectors may be controlled with adequate practices to prevent concern.

Economics are also necessary for planning and evaluating practices. Evaluation of costs and benefits is necessary to maintain agricultural viability and protect the environment.

BEST MANAGEMENT PRACTICES

Minimizing Impacts

Implementing Best Management Practices (BMPs) can minimize the potential for agricultural nonpoint source water pollution and other adverse environmental and social problems. BMPs are practices based on the best available research and scientific data. They permit efficient farming operations while achieving the least possible adverse impact upon the environment or human, animal and plant health. Selection, design and implementation of appropriate BMPs require evaluation of resources involved, and the potential impacts on them. BMPs also require evaluation of the needs for sustainable agriculture, farm operations and markets and existing practices.

Approaches to farming that seek to minimize use of agricultural chemicals and fertilizers without sacrificing economic viability are strongly recommended. These approaches are known as "Sustainable Agriculture," and "Integrated Farm Management." The goals of the various systems are to minimize chemical input and maintain environmental quality and agricultural productivity.

It is usually possible to select, combine, design and implement BMPs to protect surface and ground water and accommodate other environmental, social and economic concerns. The effects of practices on both ground and surface water quality must be considered when solving agricultural nonpoint source problems.

Infiltration of surface water may increase the potential for leaching of nutrients and chemicals into ground water.

Choosing Best Management Practices

BMPs are listed by groups for manure, agricultural compost, and chemical fertilizer in the following sections. These groups aid in preliminary consideration of appropriate BMPs for a particular farm, field, or site. Because of the diversity in farming, only the more common BMPs are listed. The type of farming and farming practices, layout, business objectives and site conditions may require that other BMPs be developed and/or used for specific applications.

Homeowners and commercial property managers need to be aware of environmentally responsible nutrient management. Lawn and landscape BMPs are provided for this audience.

Horse farms often have unique circumstances that differ from other livestock farms. Equine BMPs are listed that provide specific guidance for the growing horse industry.

Pet owners and the businesses that deal with pets also need to be concerned with animal waste management. Pet waste BMPs are designed to provide recommendations that will avoid nuisance and health concerns.

Selected BMPs should fit the operation of the entire farm and the environmental situation. The selected practices together are part of a farm plan for a particular agricultural operation. Resource professionals should select BMPs as needed and develop farm plans. Professional judgment is required to choose and implement BMPs for specific situations. The manual should not be used as a "cookbook" method to replace professional judgment.

BMP Standards

In some cases, BMPs may require standards to further define and implement them. These standards can include planning consideration guidelines and technical criteria that more specifically define what is to be done or constructed. Resource/design professionals may be needed to help choose the correct standards.

MANURE BEST MANAGEMENT PRACTICES

This group of Best Management Practices (BMPs) provides guidance for managing manure effectively, including storage, handling and utilization for forage and crop production. They give guidance in minimizing potential for surface and ground water degradation from manure use.

These BMPs are intended to permit the maximum use of nutrients and soil conditioning while achieving minimal impact upon the environment and human, animal, and plant health.

Planning Considerations

These BMPs address techniques that produce optimum forage and crop yields while limiting movement of pollutants into water bodies and ground water. Manure provides nutrients needed for plant growth. The application of manure has a beneficial influence on soil condition by improving tilth, decreasing crusting, increasing organic matter and increasing infiltration. Manure-related pollutants include nitrogen, phosphorus, pathogens, and material that has a high biochemical oxygen demand. The following steps should be taken to make maximum use of nutrients from manure:

- Obtain technical assistance from appropriate agencies to select, design, and construct or otherwise implement adequate Manure Best Management Practices. Refer to the "Some Agencies Providing Technical Assistance" section on Page 46

- Store manure in a way compatible with the type of farming operation to enhance nutrient utilization

- Determine crop production desired based on realistic yield goals

- Plan to apply manure uniformly over the maximum number of acres to avoid nutrient overload

- Keep the protection and preservation of surface and ground water in mind when performing farming operations

- Several or all of the following BMPs may be necessary to achieve the desired results. Local conditions may dictate that other BMPs be used as available

Best Management Practices

1. Control access of livestock to water bodies.

Minimize the direct deposition of manure by controlling access of livestock to water bodies. (See Appendix A)

2. Control runoff from barnyards and feedlots.

Divert clean runoff to reduce the amount of water that runs through these areas.
Control the manure-related pollutants that run off barnyards and feedlots with filter strips, grass areas below the barnyards and feedlots, and/or settling basins.

3. Divert roof runoff away from barnyards and feedlots.

Divert roof water to minimize the volume of runoff containing nutrients.

4. Manage barnyards and feedlots to minimize concentrations of manure.

Timely cleaning and removal of manure will reduce buildup, retain nutrients and prevent runoff.

5. Manage pastures to reduce concentrations of manure.

Careful placement of livestock watering facilities and herd management areas and paddock layout can reduce concentrations of manure and associated impact on water bodies.

6. Where practical, compost manure to reduce the volume of material requiring land application.

Composting converts nutrients into organic forms that are more slowly available to plants when incorporated into the soil. Leaching potential of nutrients is reduced when using compost. Composted material has little or no odor and is suitable for use as a soil amendment in residential areas. The soil structure and fertility of lawns and gardens are improved by the use of compost. Employ Agricultural Compost BMPs.

7. Store manure in properly constructed facilities or field stack during periods when land application is not suitable.

During periods when suitable sites for land application of manure are not available, the use of properly located and constructed manure storage facilities is recommended. These will provide storage until conditions permit land applications and incorporation. Field stacking is a storage alternative which requires a higher level of management than daily spreading. An intense period of labor is required to spread the stacked or stored manure. Field stacking is acceptable on flat spots away from surface water, with no direct drainage to the water.

Potential nutrient pollution of ground and surface water from improper storage or daily spreading of manure will be reduced. (See Appendix C)

8. Reduce or eliminate the use of manure in some areas.

Chemical fertilizer may be substituted for manure in some areas as a way of reducing bacterial contamination. Chemical fertilizers can be tailored to provide only those nutrients which are required by the crop. However, some nutrients from chemical fertilizers are also readily leached through the soil.

9. Maintain a balance between the number of livestock and acres of agricultural land available for spreading manure.

10. Utilize soil tests to determine background levels of nutrients and soil pH.

Amount of available nutrients in the soil reduces the need for applying extra nutrients for crop production. Over application of nutrients causes potential leaching into ground water and added expense for crop production. Proper soil pH allows better utilization of soil nutrients.

11. Base nutrient application rates on realistic yield goals.

Use crop yield and soil potential information from published county soil survey reports until yield experience information is accumulated. Only realistic goals based on recent yield experience or published soil potential information will allow accurate determination of optimum nitrogen and phosphorus application rates for crop production. Yield goal estimates should be cautiously optimistic, but not more than 10 to 20 percent above the recent average yield experienced in a particular field. It is strongly recommended that growers develop or maintain accurate recording systems for crop yield.

12. Consider nutrient contributions from legumes, other organic sources and chemical fertilizers when determining manure application rates.

13. Employ cultural practices in a timely fashion to ensure that crop yields are not depressed.

Depressed crop yields will inhibit plant fertilizer uptake. Remaining nitrogen can be leached to ground water and surface waters. Remaining phosphorus can be moved to surface waters. Proper timing of cultivation, planting, pest control, and supplemental fertilization is needed to achieve maximum crop yields.

14. Calibrate manure application equipment properly to guard against over fertilization and to achieve maximum benefit from the manure over the greatest amount of farmland.

Nutrient credits measured through manure or soil testing assume uniform and proper application. Non-uniform applications of manure result in improper nutrient crediting and shortage of manure for application on part of the farmland. This can increase the possibility of over-fertilization, which threatens ground and surface water quality.

15. Keep accurate fertilizer and manure application records and crop yield records to help determine proper manure and fertilizer rates.

Applying proper rates of manure and fertilizer can minimize risk of manure and fertilizer related pollutants to ground and surface waters. Using worksheets and keeping long-term records help predict realistic crop yield goals to plan nutrient application rates.

16. Incorporate manure applications where and when appropriate, as soon as possible after application.

This practice can reduce bacteria, organic matter and nutrient contributions from manure applications to runoff water. This, in turn limits their contribution to surface water bodies. Incorporation also eliminates odor problems. More nutrients are available for plant growth than from manure applied and left on the ground surface.

17. Avoid the application of manure on frozen ground or snow-covered fields.

Manure applications on frozen ground or snow-covered fields usually increase the amount of manure-related pollutants that reach surface water bodies.

18. Avoid applying manure directly on exposed bedrock and reduce application rates on shallow soils.

Manure should not be applied directly to exposed bedrock. Most bedrock is fractured and those fractures provide excellent pathways for nutrient migration to ground water sources. Additionally, manure application rates should be reduced on shallow soils to reflect the reduced ability to retain nutrients for plant uptake.

19. Minimize soil erosion.

Soil erosion facilitates mechanical transport of nutrients, pathogens, and organic matter to surface water bodies.

20. Diversify crop rotations to include crops that can utilize residual nitrogen where appropriate.

In sensitive areas, where nitrogen leaching may be a problem, rotating crops to include legumes or other crops, which do not require supplemental applications of nitrogen, can influence the movement of this nutrient through the soil. These crops can effectively utilize or "scavenge" any remaining nutrients left over from the previous crop or which have been mineralized from decomposing organic matter.

21. Plant cover crops on fields after harvesting annual crops, when possible.

This practice can be used in those situations where a crop is harvested early enough in the growing season to establish a cover crop. By doing this, nutrients not utilized by the primary crop can be tied up and not subject to leaching. In addition, wind and water erosion rates are decreased by the cover crop, reducing the potential for nutrient transport to surface water bodies.

22. Maintain good soil structure to reduce runoff from areas that receive manure.

Maintaining good soil structure will reduce the amount of runoff by increasing infiltration. This will reduce the potential for off-site transport of manure-related contaminants.

23. Maintain filter strips next to surface waters receiving runoff from crop fields where manure is applied.

A filter strip of perennial vegetation maintained between agricultural lands and adjoining streams and lakes will filter out some of the nutrients and contaminants before they reach the water. Minimum width of these strips can be determined by the width of any agricultural equipment used to harvest or otherwise manage the vegetation. The minimum width should be 10 feet for average slopes of less than 1 percent and proportionally up to at least 20 feet for slopes of 15 percent. Tillage should not be performed in this strip except for establishment or maintenance purposes.

24. Control spillage of manure when transporting from the storage area to the field.

Manure along roadways presents a nuisance and safety problem especially near urban areas. Runoff from roadways can carry the spilled material into streams causing water quality problems. Use hauling equipment consistent with the type of manure generated. Limit passage of hauling equipment through the manure as much as possible to reduce tracking on roads. Effort should be made to clean up excess amounts of manure deposited on public roads.

25. Manage milk house and parlor wash water to avoid migration of nutrients to brooks, streams and lakes.

Wash water carrying nutrients and manure can be disposed of using land application, filter strips, constructed wetlands, organic matter beds, lagoons, subsurface disposal, storage structures, or by mixing with manure for land application.

26. Manage manure to control excessive fly populations.

Manure storages are often blamed for fly problems. However, manure that is stored and has a dry, crusted layer isn't a primary fly-breeding area. Manure can also be contained within the housing system with slatted floors or a bedded pack. These are not primary fly-breeding areas as long as the top layers are undisturbed. The key is to avoid wet, rutted areas around storage facilities where flies will breed. Impermeable containment walls, proper drainage, or catch basins may be needed to avoid pockets of water stagnation. Chemical pesticide control may be necessary.

27. Control odors as necessary.

Movement and distribution of manure will create odors even under the best practices. Every effort should be made to minimize odor problems from daily operations by: (a) keeping neighbors informed of activities, (b) condensing spreading time in each field to minimize periods when odors can be most offensive, (3) incorporating manure immediately on tilled fields, and (d) utilizing vegetative barriers that can buffer odor drift.

MANURE IRRIGATION BEST MANAGEMENT PRACTICES

The following Best Management Practices (BMPs) for Liquid Manure storage, transportation and application are to be used in addition to and in conjunction with those BMPs published in this manual.

Liquid manure systems provide advantages and disadvantages to the agricultural operator. Due to the high water content of the material, the volume of material that must be stored, transported and land applied is far greater than is the case with solid or semi-solid manure management systems. In order to efficiently utilize the material with the least possible adverse effect on the environment and to minimize nuisance-type problems, operators should be sure that storages, conveyances and application equipment are properly constructed, maintained and calibrated.

Best Management Practices

1. **Have samples of liquid manure analyzed at least annually. At a minimum the following constituents should be determined:**
   - **% Total Solids**
   - **Total Nitrogen**
   - **NH_3N Nitrogen**
   - **Phosphorous (P_2O_5)**
   - **Potassium (K_20)**

   It is preferable that samples be analyzed shortly before land application is to begin. N losses in open storages can be significant. The percent solids of liquid manure typically will be less than 4% and should not exceed 8%.

2. **Maintain tanks used for transporting liquid manure from storage areas to application sites to prevent leakage onto public roadways.**

3. **Pits or lagoons used for short-term storage during the course of a field application activity should be lined with clay or a suitable impervious material. Always allow freeboard in the pit to accommodate rainfall so as to avoid overflows.**

4. **Check pumps, pipes, hoses and nozzles at least daily for signs of leakage. Repair damage promptly to prevent over application or ponding in spots. If underground pipelines are used, they should be carefully assembled and tested for leaks.**

   Because of the corrosivity of liquid manure, underground delivery systems should be constructed of plastic or non-corrosive materials. Flushing lines with water will help prevent blockages from occurring and extend the life of the equipment.

5. **Avoid irrigation with liquid manure when the soil is saturated or when excessive rainfall causes ponding or runoff. When the water table is within 6" of the surface, irrigation should be delayed.**

   Irrigation with manure/milkroom waste provides both plant nutrients for growing crops and necessary water. Under normal circumstances, application rates will be based upon the nitrogen needs of the crop. However, there may be occasions when an excess of applied water may become the limiting factor in determining the maximum application rate.

6. **Inspect fields for broken tiles and other possible short-circuit routes that could result in a direct discharge of manure to drainage tile and surface-drainage ditches.**

   Avoid or limit liquid manure applications in areas where conditions would cause a discharge to occur.

7. **Determine hourly application rates and match the applied volume of material to the infiltration rate and permeability of the soil.**

   Lower application rates and multiple passes with irrigation equipment may be required to prevent runoff and ponding.

8. **Limit an application event to an amount that will bring the soil to field moisture capacity.**

   Field moisture capacity is the amount of water a given soil will hold following saturation and after the force of gravity has drained all the water it can.

9. **Limit the annual application rate so as not to exceed the crop's annual nutrient requirements.**

   Take into account N losses to the atmosphere during application. If necessary, consult with Cooperative Extension or qualified crop consultants for the nutrient requirements of particular crops.

10. **Monitor wind conditions throughout the day to assure that over spray or drift onto surrounding properties, public roadways, surface water bodies or environmentally sensitive areas does not occur.**

    Avoiding irrigation during high temperature and/or high winds will minimize N losses as well as minimizing potential odor problems. When irrigating with livestock manure, the operator should be aware of odor nuisances that may affect neighbors. Spray irrigation produces aerosol sprays that can be detected for long distances. Wind direction and impact on neighbors need to be observed closely.

Horse Facility BMPs

For purposes of these BMPs "horse" refers to all members of the equidae family which may include horses, asses, mules, zebras, ponies or donkeys. These BMPs may also be applied to members of the camelidae family including llamas, alpacas, quanacos, vicunas and camels; to members of the cervidae family including deer and elk; and to other domestic animals kept and managed similarly to horses.

Horse-keeping facilities share some concerns with other livestock farming operations, particularly the need to manage large volumes of manure. Many equine operations lack cropland for utilizing manure at agronomic rates. Agronomic rates are those at which growing crops can utilize the nutrients present in the material. If the number of animals exceeds the ability of the available cropland to utilize the nutrients present in their manure, provisions for other means of use or disposal must be made.

Horse manure, like the excreta of other animals, has the potential to carry and transmit human and animal disease. Horse manure also contains significant levels of nitrogen and phosphorous - nutrients that can lead to water quality problems of concern to public health and the environment. On the other hand, horse manure, if properly managed, can be an asset to the production of crops as well as for horticultural and landscape uses.

Horse manure, unlike dairy or poultry manures, is usually mixed with a large proportion of bedding - generally sawdust or wood shavings and uneaten or spoiled hay. Composting and field application considerations are also quite different with horse manure than with concentrated types of manure. Due to the high proportion of bedding material in stable waste, application directly to cropland may in the short term do more harm than good to crops by tying up nitrogen needed to decompose the wood residue; making nitrogen unavailable to the crop.

Horse farms sometimes are located on steep, rocky, wet, or otherwise challenging terrain that severely limits site choices for manure storages or for the development of pasture or cropland. Ideally, nutrient management should be considered well in advance of the establishment of an equine operation. Barns, riding arenas, exercise yards, paddocks, and manure storage sites need to be sited with an eye to managing nutrient runoff as well as for practical reasons relating to running an efficient operation.

Selecting locations for paddocks, trails, and exercise areas need to be made with careful consideration for topography and drainage. Paddocks that are wet, muddy, and manure-laden can lead to problems with horse hoof health. Steep slopes and icy build up may also result in serious injury to horses and/or riders.

A key consideration in the establishment of a horse-related operation – whether a business or a single horse for personal use - should be the suitability of the site for maintaining adequate exercise and grazing areas for the animals as well as for managing manure. Both objectives need to focus on the maintenance of water quality. Possible nuisance issues such as flies and odors also need to be analyzed.

Best Management Practices

1. Properties that board equines should have a properly sited manure storage area or structure.

The most important consideration for siting a manure storage should be the site's avoidance of areas where surface water (streams, rivers, lakes, ponds or wetlands) is present or where runoff from the land, building roofs, or other surfaces would transport nutrients from animal waste to surface water. Proximity of the manure storage to stables is a practical consideration but should not be the primary consideration in siting a manure storage. (See Appendix C)

2. Manure storages should be designed in a manner that considers the site's potential to cause environmental degradation.

Field stacking on level ground may be adequate in some situations provided there is no direct runoff to surface water. Field stacking must be outside of the protective radius around drinking water wells (75 feet for private wells; up to 400 feet for public water supply wells). Where field stacking is suitable, the site should be alternated periodically so as not to create nutrient "hot spots" – places where nitrogen and phosphate build up to excessive levels.

3. Where properly sited field stacking is not an option, an adequately constructed and sized storage structure should be constructed. A well-designed and engineered manure storage structure should be considered a pollution control facility.

In New Hampshire, manure storages need to provide adequate capacity for approximately five or six months of storage during winter when land application is not appropriate and removal from the site may be difficult. USDA NRCS, the County Conservation Districts, or UNH Cooperative Extension can be consulted for storage capacity calculations and construction methods for storage structures. In most cases, a three-sided storage structure on a concrete pad is preferred. Sides can be made from concrete blocks or heavy planks. When challenging environmental conditions are present on a site, for example - steep slopes, a nearby well, or compacted soils are present - a roof is also desirable. Roofing the storage will nearly eliminate the potential for contaminants leaching from the storage.

Size your storage facility according to the number of animals and the number of days you intend to hold the material before use or removal. To estimate the minimum base area of your storage use the following formula:

- Number of Animal Units (AU equivalent to 1,000 lbs.) One AU produces 1 cubic foot of manure per day. Multiply 1 cubic foot x number of AU x the number of days storage needed = cubic feet of manure.
- Number of AU x cu. feet of bedding per day (use 1 cu. foot if unsure) x number of days storage needed = cu. feet of bedding
- Cu. feet of manure + cu. feet of bedding = total volume in cubic feet

- Sq. feet of area required = total volume divided by desired maximum average height of manure and bedding in the structure.

EXAMPLE: Lets say you have 3 horses for which you need to provide 6 months of storage between removal.

3 horses x 1,200 lbs. each = 3,600 lbs./1,000 lbs. per AU = 3.6 AU
3.6 AU x 1 cubic foot of manure per day x 180 days = 648 cubic feet of manure
3.6 AU x 1 cubic foot of bedding per day x 180 days = 648 cubic feet of bedding
648 + 648 = 1296 cubic feet of manure & bedding
1296 divided by 6 feet (desired maximum average height of storage) = 216 square foot storage area needed. A 12 foot x 18 foot area (or equivalent dimensions) is necessary. (See Appendix B)

4. Manure storages, whether a temporary site or permanent structure, should not be the final resting place for manure.

A fundamental principle of BMPs is that manure is a resource. Accordingly, manure should be removed periodically and used for its soil amending properties. However, if suitable crop or other land application sites are not available horse manure needs to be properly disposed of as other solid wastes. *A preferred alternative to disposing of manure as solid waste, is to transport the material to a composting site where it can be recycled into a beneficial soil amendment product. Some municipalities maintain town composting facilities for residents. Commercial composters and other farmers with suitable sites who recycle organic materials as a business are also a possible outlet for manure. Commercial composters or waste disposal firms may also provide dumpsters for short-term storage and/or transportation services for manure.*

5. Composting horse manure and bedding can yield an end product with desirable soil amending properties for horticultural applications. Composting will also significantly reduce the volume of material that will need to be handled.

In order to successfully compost horse manure and bedding, a high level of management is required. Frequent turning or some other method of aerating the pile is important for rapid composting. UNH Cooperative Extension and USDA NRCS can provide detailed guidance on composting methods.

6. Pastures, paddocks, and adjacent buffer areas need to be carefully managed to minimize the erosion and sedimentation that may result from water flowing over bare ground. Erosion may carry soil particles, nutrients, and disease causing organisms to waterways.

Fencing needs to be designed to allow for rotation of grazing areas on properties with limited land area. Maintain a vegetated buffer between heavily used areas and surface waters. Manage drainage to keep it from becoming concentrated as it flows through the buffer. In order for a vegetated buffer to protect water quality, runoff must flow evenly across the area.

7. Establish a Sacrifice Area. "Sacrifice Area" is a selected area set aside or rested from the grazing system and used to confine animals in order to protect pastures from over-use at critical times such as during winter months, mud season, and times of slow pasture regrowth, for example, during drought. Creation of a "sacrifice area" will also permit other areas of pasture adequate time to recover from heavy grazing pressure or traffic.

Care should be taken to select a sacrifice area that is on nearly level ground and not prone to runoff from buildings or hardened surfaces, and separated from surface water by a vegetated buffer (see 6 above). An ideal slope of 2-4% will reduce ponding of water and will have less erosion than steeper slopes. (A slope of 2% means that over a distance of 100 feet, the elevation will change by 2'). A slope of 1-7% may be adequate. A well-planned sacrifice area should be the minimum size necessary to comfortably accommodate the number of animals present. The best surface for a sacrifice area may be one on which the topsoil has been removed and replaced with 8"-12" of well-drained gravel with an assortment of particle sizes from sand and some fines up to stones of one inch in diameter. A surface covering of stone dust or sand is gentle on hooves. Some spots may require subsurface drainage where a high water table exists. Particularly wet spots will benefit from the use of geotextile fabric covered with 8+ inches of gravel base and your choice of surface material.

8. Equine operators must practice regular or in some cases daily pick-up of manure from concentrated sacrifice areas and denuded paddocks. When manure is trampled into bare ground the organic matter in the waste holds far more water than well-drained sand and gravel, thus making the site more and more muddy as time goes on.

Providing well-drained material for sacrifice areas, exercise areas and paddocks improves footing, reduces ponding, and minimizes the presence of flies and parasites. Manure and manure/bedding mixtures should never be used for fill.

9. When streams, ponds, springs, or wetlands exist on a site, access of horses should be controlled by the use of fencing or other means. Controlled access means that the ability of animals to cross or enter wet areas should be limited to allowing them access for drinking, or for crossing at a particular point. In some situations, such as the presence of downstream human drinking water sources, where public swimming occurs, or where flowage drains directly to a lake, access of livestock to streams should be eliminated.

The access point for livestock drinking water sites should be carefully selected to minimize the erosion and sedimentation that might occur on steep or muddy sites. Fencing should be configured to permit only enough access for the head and forefeet of the animal. Similarly, when wetland crossings are necessary, the crossing point should be at a place where hoof action will cause the least damage to banks. USDA NRCS can provide detailed plans for fencing and controlled crossing construction.

10. When water is available to animals from other sources and access to surface water drinking sites is not necessary, fencing should exclude animals from water.

Stock tanks, fountains, frost-free waterers, and pasture pump devices can provide alternatives to drinking from surface water. USDA NRCS, County Conservation Districts, and UNH Cooperative Extension can provide guidance on selection and use of these devices.

Pet Waste BMPs

Pet wastes are often not thought of as sources of water pollution. However, pet droppings, like livestock manure, contain nutrients that can contribute to nonpoint source pollution or polluted runoff. In addition, pet wastes can harbor a number of disease organisms that are transmissible to humans either by water-borne means, through soil-contact transmission or by direct contact with the waste.

More than half of all New Hampshire households own at least one dog, cat, or other pet. By properly managing pet waste, citizens can avoid contributing to water quality impairment, minimize odors and fly problems, and protect the health of family members, neighbors and other pets.

Diseases that can be transmitted from pet waste to humans include:

Campylobacteriosis – A bacterial infection carried by dogs and cats that causes diarrhea in humans.

Cryptosporidiosis - A protozoan infection that causes diarrhea and abdominal pain.

Giardiasis – A protozoan infection of the small intestine that can cause diarrhea, cramping, fatigue, and weight loss.

Salmonellosis – The most common bacterial infection transmitted to humans by other animals. Symptoms include fever, muscle aches, headache, vomiting, and diarrhea.

Toxocariasis – Roundworms usually transmitted from dogs to humans, often without noticeable symptoms, but in humans can cause vision loss, rash, fever, or cough.

Toxoplasmosis – A protozoan parasite carried by cats that can cause birth defects such as mental retardation and blindness if a woman becomes infected during pregnancy. It is also a problem for people with depressed immune systems. Symptoms include headache, muscle aches, and lymph node enlargement.

Best Management Practices

1. Do not use pet wastes, either composted or uncomposted, in or near crops intended for human consumption.

Due to the variety of pathogenic organisms that can transmit illness to humans, the risk from using pet wastes on food crops is too high to justify its use.

2. If properly composted in an area not used for other composting, pet waste and bedding may be suitable for use on ornamental crops, turf, or forestry uses in areas with a low likelihood of human contact. Use of pet waste, even composted pet waste, should be

avoided in any areas where children would likely play.

Pet manure contains on average twice the nitrogen, about the same phosphorous, and half the potassium as cow manure. If the use of composted pet waste on non-food crops is considered, the rate of application should be based on the results of testing both the compost and the soil. Composting pet wastes and bedding can reduce the presence of pathogens and parasites if composting temperatures of at least 165 degrees F for five days can be maintained and with frequent turning to expose the outer portions of the pile. **Small home compost piles are rarely able to reach the high temperatures needed to kill pathogens and therefore the incorporation of pet waste into home compost piles is not recommended.**

3. Homeowners with just a few pets usually are best off to dispose of the material with other household trash. If adequate yard space is available, small quantities of pet waste may be buried.

If burial is chosen as a disposal method, choose a site at least 100 feet from wells, surface water or areas prone to runoff. Dig a hole or trench at least five inches deep, cover with soil, and tamp. Soil microorganisms will breakdown the waste in several months. Alternate burial sites at least once a year. Commercially available in-ground pet waste digesters may also be suitable for disposing of the wastes of less than four animals. Digester systems utilize enzymes and bacterial action to break down wastes.

4. It is necessary that commercial kennels, breeding facilities, dog race tracks, pet day care facilities, and pet shops have a plan for short-term storage and regular removal of pet wastes from the premises.

Generally, commercial pet facilities should be incorporating pet waste into properly sized private septic systems or into the municipal wastewater treatment system, if available. In some cases, dry manure/bedding management systems may be suitable. Provided adequate land area is present, a well-made storage structure with an impermeable base may be used to store and/or compost pet waste for a time prior to disposal.

5. Commercial pet facilities that have outdoor runs or pens should pick up droppings on a daily basis and incorporate them into the preferred disposal method. Outdoor runs and pens should be sited on flat areas and avoid areas subject to surface water runoff. If runoff occurs, gutters and downspouts should be installed on buildings to minimize runoff through pens, runs, and exercise areas.

If outdoor pens and runs are roofed, the likelihood of waste runoff is greatly reduced. Diversions should be employed to keep rainwater from moving through exercise areas and runs. Berms or collection systems should be utilized to redirect contaminated runoff away from surface water or wellhead areas.

AGRICULTURAL COMPOST BEST MANAGEMENT PRACTICES

This group of Best Management Practices (BMPs) provides guidance for implementing agricultural composting. Composting is the aerobic biological decomposition of organic matter including manure, leaves, bedding and crop residue. It is a natural process that is enhanced and accelerated by the mixing of organic waste for optimal microbial growth to produce a relatively stable soil amendment.

These BMPs are intended to permit the maximum use of nutrients and soil conditioning while achieving the least possible impact upon the environment and human, animal, and plant health. These BMPs give guidance in minimizing runoff and leaching into the surface and ground waters and risk to water quality from nutrient enrichment.

Planning Considerations

Most of the nutrients in agricultural compost are in a stable organic form, which are slowly released to growing plants. Only 8-12 percent total nitrogen (N) is available the first year following its application. Nutrients from chemical fertilizer are nearly 100 percent available to growing plants.

Benefits of agricultural composting include:

- A lowered risk of pollution by stabilizing nitrogen in an organic form, and reducing its loss to ground and surface water
- Improved handling
- A saleable product that is a good soil conditioner
- A lower risk of nuisance complaints
- Destruction of weed seed

Disadvantages of agricultural composting include:

- Labor and equipment costs
- Weather delays and odor
- Marketing considerations
- Diversion of manure and crop residue from cropland
- Slow release of nutrients
- Risk of losing farm classification if composting extends beyond using normal farming materials

Several or all of the following BMPs may be necessary to achieve the desired results. Local conditions may dictate other BMPs be used as available.

Best Management Practices

1. Determine the method of active composting that best fits the operation based on realistic labor, equipment cost, and availability.

The three methods of active composting include: windrow, static pile, and invessel. The windrow method is the strategy most commonly used to produce agricultural compost.

2. Evaluate facility and application sites for environmental constraints.

Constraints include soil and surface drainage, depth of bedrock, setbacks and prevailing winds.

3. Analyze compost constituents for developing the best recipe for the mix and monitor the nutrient content.

4. Determine manner of compost utilization.

If cropland is available, employ Manure BMPs that pertain to crops, soil, and application techniques. If compost must be removed from the farm, develop marketing or give-away program.

5. Plan timing of compost application to avoid periods when soil is frozen.

6. Determine realistic yield goals.

Consider soil potential, climate, and management.

7. Base nutrient application rates on realistic yield goals.

8. Consider nutrient contributions from legumes, other organic sources, and chemical fertilizers in determining application rates.

9. Calibrate compost spreading equipment to guard against over-application.

10. Store compost in properly constructed facilities during periods when land application is not suitable.

Field stacking is acceptable on flat spots away from surface water.

11. Avoid spillage during handling and transportation to minimize nuisance and safety problems.

12. Establish a practice of routine soil testing and record keeping for residual nutrients.

Consider nutrient credits from other sources, and make effective use of presidedress nitrate-N test (PSNT), which is unique to corn crops.

13. Control vectors, including flies, and odors to prevent public health problems and nuisance.

Livestock Mortality Composting BMPs

Farmers faced with the disposal of livestock mortality have limited choices. In the past, rendering services provided free or low-cost pick-up services for farm mortalities. Recently, end market prices for rendered livestock by-products have declined markedly due to Bovine Spongiform Encephalopathy (BSE) concerns and other factors. Where dead stock pick-up service is available, costs to farmers for the service are rising. Currently, there are no rendering companies that provide pick-up service for dead stock from farms in New Hampshire.

Traditionally, burial and natural scavenging (the "back 40 method") have been used for disposing of mortalities on farms. Burial is an option that has several disadvantages. Deep burial may affect groundwater and wells while surface burial may impact surface water and attract scavenging animals. Shallow or deep burial may not be advisable at all in some areas due to shallow depth to water table, to bedrock, or to the proximity of nearby surface waters.

The practice of placing livestock mortalities where wild animals may scavenge them also presents problems. Scavengers may become habituated animals and may linger around the farm waiting for the next feeding, or they may prey on live animals. Scavengers can harbor and transmit diseases to farm animals. In more suburban locations, domestic dogs can cause problems by bringing home remains of carcasses. In many places, public perceptions of the "natural disposal method" may not be favorable, leading to nuisance complaints.

Composting is a natural process of decay that can prevent problems with water quality, odors, flies and scavengers. The finished compost can then be used beneficially on growing crops. Composting animals on-site reduces biosecurity concerns. Trucks coming to and going from the farm, from other farms, especially if carrying mortalities, present an obvious biosecurity risk.

The New Hampshire Department of Environmental Services (NHDES) requires a permit for certain types of composting facilities. However, NHDES does not require a permit for farms that compost their own farm's mortality on-farm, provided it is done in accordance with these BMPs. If you compost animals belonging to another person, or another person composts your animals, then a permit is required from NHDES. Composting, when limited to manure, crop residues, and leaf and yard wastes, does not require a permit, whether inputs are brought in from off site or not.

Livestock mortality composting is quite different from other types of composting as it is a static pile process. The pile should not be turned until the process is substantially complete. At this time, composting of butchering wastes and meat-based food waste are not subjects covered by this BMP. Mass composting of farm livestock as a method of disposal for emergency animal disease control depopulation programs, is also not covered by this BMP.

Best Management Practices

1. Select a composting site that is dry, well drained, and slightly sloped (1-2% best), with good all-weather access.

Depending on slope and vegetative cover the site should be:

- At least 100 feet and up to 300 feet from any surface water;
- At least 200 feet and up to 400 feet from a wellhead;
- Chosen to avoid areas with exposed or shallow distance to bedrock;
- *Note: If the composting site is located on a concrete or impermeable pad, or if the composting site is roofed over, or if the composting takes place in a self contained enclosed vessel and is adequately protected from leaching or runoff of liquids, the above distances can be reduced. Remember that a sufficient layer (2-3') of absorbent bulking material under the carcasses is meant to absorb any liquid and prevent runoff and leaching.*
- At least 3 feet above the seasonal high water mark
- At least 150 feet from inhabited buildings and on abutting property;
- At least 50 feet from public roads.

2. Maintain an ample supply of one or more suitable bulking agents in which to encase the carcass.

A full-grown cow requires 12 cubic yards of bulking materials such as woodchips, shavings, sawdust, chopped straw, silage, dry manure or finished compost. Begin by laying down a base of no less than 2 feet of bulking agent. Place the carcass on its side in the center of the base and cover with at least of feet of bulking agent on all sides and on top. Young cattle and small animals can be lined up in a row and layered with 2 feet of bulking agent between layers.

3. Monitor the composting process by keeping a logbook.

Record such factors as the starting date, the type and volume of bulking agent used, and the temperature inside the pile. In order to assure the destruction of most pathogens and weeds, an internal temperature above 131 degrees F should be reached and maintained for at least three days.

4. Allow the pile to sit for 4-6 months without turning.

At the end of 4 months begin probing the pile with a bucket loader to inspect for thorough decomposition. If soft tissues remain or strong odors are released, turn the pile, recover with additional bulking agent and wait another two months. If only large bones remain, remove them for return to the next compost batch.

5. Finished compost can now be used for land application at agronomic rates or reused in composting.

Finished livestock mortality compost may be spread on crop land for incorporation into the soil. Due to uncertainty regarding BSE concerns, it is not recommended that livestock mortality compost be used on human food crops. Compost produced from ruminant animals should not be used on land pastured by ruminant animals or on hay crops intended for consumption by ruminant animals. Ruminant animals include cattle, sheep, and goats.

CHEMICAL FERTILIZER BEST MANAGEMENT PRACTICES

This group of Best Management Practices (BMPs) provides guidance for managing chemical fertilizer use to minimize nutrient runoff and leaching into surface and ground waters. These practices provide guidance for the proper use of chemical fertilizers for commercial agriculture, and other commercial applications, parks, cemeteries and recreation and other areas. They also provide guidance to commercial applicators and others for home lawns and gardens.

These BMPs are intended to permit the maximum use of nutrients and soil conditioning while achieving the least possible impact upon the environment or human, animal and plant health. They will reduce the potential for ground and surface water nutrient contamination, increase the efficiency of fertilizer use, and educate users about the proper use of chemical fertilizers.

Planning Considerations

Proper planning is required when choosing BMPs for chemical fertilizer use for a particular farm, field or site. Several or all of the following BMPs may be necessary to achieve the desired results. Local conditions may dictate that other BMPs be used as appropriate.

Best Management Practices

1. Determine the surface and ground water contamination potential of each site.

Consider soil type, slope, depth to bedrock or impervious layer, and location and depth or distance to aquifers and bodies of water.

2. Assess actual plant (crop) needs.

Realistic yield goals have been established for selected plants for each soil map unit in each county. Data is available from the Natural Resources Conservation Service at each county field office.

3. Utilize soil tests to determine current nutrient levels and soil pH.

The standard soil test will measure soil levels of phosphorus, potassium, calcium and magnesium as well as soil pH. Nitrogen, which the standard soil test does not measure, and phosphorus are the two nutrients of greatest concern as potential water pollutants. Use soil pH testing and adjust soil pH to appropriate levels which can reduce nutrient loss significantly, while increasing plant growth and yields. Soil tests should be coupled with crop nutrient need (based on realistic plant yield goals) when determining fertilizer application rates.

Soil nitrate testing (PSNT) is now available for use in determining supplemental nitrogen needs for corn. A typical fertilizer program for silage corn would include the pre-plant incorporation of 15-20 tons of dairy manure per acre, the use of a starter fertilizer at planting, and the application of supplemental nitrogen at the 8 to 16 inch stage of plant growth based on need as determined by soil nitrate testing.

4. Assess all available nutrients including manure and other organic sources and legume contributions.

5. Apply nutrients only at levels required for plant growth.

Follow fertilizer recommendations such as made by UNH Cooperative Extension are designed to provide maximum economic return with the least impact on the environment. The UNH Cooperative Extension provides detailed fertility management guidelines and fertilizer recommendations for all crop and plant systems. These recommendations are based on crop need, soil type and condition, and soil or plant tissue analysis, or both, and include fertilizer type, rates and timing.

6. Use realistic plant growth expectations and appropriate timing for application of chemical fertilizer.

Nutrient applications should be timed to coincide with periods of maximum plant or crop need. The application of several smaller amounts of fertilizer timed to coincide with plant need will generally require less total fertilizer than loading the soil with an early season or pre-plant single application, thereby reducing the potential for nutrient loss to surface and ground waters.

The following is an example of current UNH Cooperative Extension recommendations. With strawberries, a small initial nitrogen application is made at planting, with additional small nitrogen applications at runner initiation and runner rooting. This regimen can reduce the total nitrogen applied by almost 50 percent when compared to a single, pre-plant application.

7. Use split fertilizer applications where possible. Using smaller applications on a more frequent basis will decrease potential for nutrient loss to ground or surface waters.

Home lawns, depending on the quality of turf desired, may receive between one and three applications of fertilizer annually with three applications being the maximum for most situations. A good guideline for a three application schedule includes use of a starter type fertilizer in May (1-2-1 ratio), a slow release high nitrogen fertilizer (having a 4-1-2 ratio) in July, and balanced fertilizer (1-1-1 ratio) in September. If a single application is to be applied, the September application is best.

8. Develop plant management systems that maintain soil organic matter levels at a minimum of three percent to five percent to improve soil nutrient retention.

9. Apply nutrients uniformly.

10. Use less leachable forms of fertilizer (slow release) where possible.

11. Use mulches, both organic and synthetic, to significantly reduce the risk of leaching nutrients from the soil, and to reduce the total amount of chemical fertilizer applied to achieve optimum growth and yields.

For example, plastic mulches are often used in vegetable crops such as vine crops, tomatoes and peppers. Fertilizer is applied pre-plant under the mulch which prevents leaching of nutrients from the root zone, increasing efficiency of use and reducing risk of off-target movement.

12. Avoid applying nutrients to very shallow soils or exposed bedrock.

13. Calibrate chemical fertilizer application equipment properly to insure accurate application.

14. Keep accurate fertilizer and manure application records and crop yield/plant growth records to aid in crop/nutrient management.

15. Schedule irrigation to minimize leaching potential (avoid excessive irrigation).

16. Diversify crop rotations to include crops that can utilize residual, leachable soil nutrients.

17. Manage fertigation systems so that nutrients are incorporated into irrigation water only when crops require supplemental fertilizer applications.

Fertigation should not be used to replace an integrated fertility management program-which includes programming the use of starter fertilizer, manure and other organic sources, and split fertilizer applications, all based on soil and/or tissue analysis and plant need.

18. Use plant tissue testing to determine potential or existing macro and micro nutrient problems.

Use recommendations such as provided by UNH Cooperative Extension. Plant tissue testing is an excellent tool for determining exact plant nutrient needs for many essential plant nutrients, including nitrogen, phosphorus, potassium, calcium, magnesium, iron, zinc, manganese, copper, boron, molybdenum, chlorine, sulphur, and others. It is routinely used for fruit crops.

19. Limit applications of nitrogen fertilizers to coincide with plant uptake.

Nitrogen fertilizer applications should be timed to coincide with plant uptake. Spring applications should be when annual crops are planted or when biennial or perennial plant growth begins. Fall applications should not be made after plant growth has ceased or when the additional fertilizer will no longer have an impact on yield.

20. Do not apply nutrients during winter months when ground is frozen or snowcovered because of the high risk of runoff.

21. Minimize soil erosion.

Nutrients, in particular phosphorus, are often attached to soil particles and can be transported with the soil particles as they are washed away by the erosion process.

22. Employ pest control practices such as recommended by UNH Cooperative Extension to insure that plant growth is not depressed.

23. Plant cover crops after harvest of annual crops to minimize soil erosion and provide soil organic matter.

Cover crops also tie up nutrients that may otherwise be leached from the soil or runoff, making them available to future crops.

24. Install filter strips next to surface waters receiving runoff from areas to which fertilizers have been applied.

A filter strip of perennial vegetation maintained between agricultural lands and adjoining streams and lakes will filter out some of the nutrients and contaminants before they reach the water. Minimum width of these strips can be determined by the width of any agricultural equipment used to harvest or otherwise manage the vegetation. The minimum width should be 10 feet for average slopes of less than one percent and proportionally up to at least 20 feet for slopes of 15 percent. Tillage should not be performed in this strip except for establishment or maintenance purposes. Artificial wetlands and/or basins can serve as efficient nutrient traps.

25. Use leguminous rotation crops to reduce the need for chemical fertilizers.

26. Store fluid fertilizers in labeled containers and/or structures that prevent the discharge of fluid fertilizers and are resistant to corrosion, puncture, or cracking.

27. Store and handle dry fertilizers in a manner to prevent pollution by minimizing losses to the air, surface water, ground water, or subsoil.

Lawn Care/Turf & Landscape Fertilization BMPs

Turf makes up a large part of the urban and suburban landscape. Home lawns, golf courses, parks, cemeteries, and commercial developments all utilize turf to create a pleasing environment. A properly managed lawn area can provide a net benefit to the environment. A thick, healthy sod can filter and absorb nutrients and pollutants. On the other hand, improper fertilization practices on lawn and landscaped areas can contribute excess nutrients to surface and ground water.

Nitrogen contained in fertilizers can be a source of ground water contamination. Nitrogen and phosphorous can be factors in polluting lakes and ponds. Excessive nitrogen in drinking water is a public health concern while both nitrogen and phosphorous contribute to eutrophication of surface waters – the proliferation of aquatic plants and algae. Potassium, the third major plant nutrient is not considered an environmental or public health concern.

Best Management Practices

1. Test turf soil every two years to determine existing nutrient levels and pH. The University of New Hampshire Cooperative Extension office in each county can provide information on how to sample soils and will interpret laboratory results. (See the listing of UNH Cooperative Extension offices elsewhere in this publication.)

Soil acidity or alkalinity is measured as "pH" with 7 being neutral. Most turf grass species grow and utilize nutrients best at a pH between 6.5 (slightly acidic) and 7. Ground limestone is used to raise the pH of acid soils to neutrality.

2. When soil test results call for the addition of fertilizer, a commercial fertilizer should be chosen that most closely matches the recommendation. Learn to understand the terminology used on fertilizer labels.

All commercial fertilizer labels are required by state fertilizer law to provide certain information to consumers in the form of a "Guaranteed Analysis" statement. Three numbers are used to indicate the "grade" of a fertilizer product, for example "23-3-6." The grade represents the percentage by weight of "Total Nitrogen (N), Available Phosphate (P2O5) and Soluble Potash (K2O)" contained in the package. The nitrogen guarantee may be further broken down into the various forms of N. (See Page 39 for an explanation of the terms on a fertilizer label.)

3. Sources of fertilizer nitrogen can be either readily available (soluble) or slow or controlled release. Fertilizer particles may also be coated to provide another method of controlled release of nutrients. Utilize slow or controlled release nitrogen sources, which become available to plants gradually. If highly soluble fertilizer sources are used they should be applied in several smaller split applications.

The nitrate form (NO_3) of nitrogen (N) is readily available to plants but also leaches freely. The ammoniacal form of nitrogen (NH_4) tends to chemically bind with soil particles and provides a more controlled release of N to plants with less leaching. Urea is a highly soluble source of nitrogen while ammonium nitrate is less soluble. Other nitrogen source materials fall somewhere in the middle range for solubility. Sulfur or polymer coated urea products provide a less soluble means of employing urea that is better for the environment.

4. Phosphate (P_2O_5) is not particularly soluble in the soil and slow release forms are not available commercially. Phosphate application on established lawns is usually only needed at low rates. Phosphate should be avoided entirely when soil test results for P are high or when steep slopes would carry particles to nearby surface water.

A few fertilizer products that contain coated phosphate carriers are starting to enter the marketplace. A "No P" fertilizer would have a grade like this: "18-0-3."

5. Time fertilizer applications in conjunction with rainfall or irrigation. If heavy rain is anticipated do not fertilize as nutrients will be flushed from the lawn into drain ways and low areas.

Ideally, a fertilizer application should be irrigated with no more than ¼" of water, provided the soil is not already saturated. The amount of applied water can be measured with a rain gauge.

6. Before applying fertilizer to an area, measure the area to be fertilized to determine the square footage of the area.

If the lawn is a simple square or rectangle, measure the length and the width, multiply the two figures to arrive at the total square footage. If the lawn is an irregular shape, approximate a square or rectangle and use these figures to calculate the square footage. Almost all lawns can be divided into a number of smaller squares and rectangles that can be measured individually. Add the square footage of all the smaller pieces to arrive at the total square footage to be fertilized. Bring this figure to the store to determine how many bags of fertilizer you will need to cover the entire area.

7. When using dry granular fertilizers adjust the spreader setting to match the desired rate and fertilizer spreader model shown on most turf fertilizer bag instructions. A more precise method of gauging the actual application rate can be determined by calibrating the spreader.

Spread a 10-foot by 10-foot piece of plastic sheeting or canvas on the lawn and pass the spreader over the area at a given spreader setting. Then collect the fertilizer on the sheeting and weigh it to determine the actual application for 100 square feet. This can be compared to the recommended rate which the label usually states in pounds per 1,000 square feet. Record the rate actually applied for the particular spreader setting. Remember that this rate will only be correct for the same brand and formulation of fertilizer.

8. Avoid applying fertilizer within 10 feet of culverts, drainage ditches, wells, roadways and walks. If fertilizer is spread on driveways or walkways clean it up with a broom. Pick up any fertilizer spilled on the lawn as well.

The New Hampshire Comprehensive Shoreland Protection Act, RSA 483-B:9, prohibits the application on lawn and turf of any fertilizer within 25 feet of most lakes and streams and requires the use of low phosphate, controlled release fertilizer only within 250 feet of the high water mark of lakes and streams.

Fertilizer Labeling

A sample label for 12-2-8 fertilizer might show:

JONES FERTILIZER COMPANY 12-2-8

Guaranteed Analysis:

Total Nitrogen (N)	12%
Water Insoluble Nitrogen (WIN)	6%
Available Phosphate (P_2O_5)	2%
Soluble Potassium (K_2O)	8%

Net Weight 50 lbs.

New Hampshire has a law governing the labeling and sale of fertilizers. The manufacturer must place a statement guaranteeing the weight and analysis of the fertilizer on, or attached to, the bag or container. If the fertilizer is delivered in bulk, a written statement containing the same information must be supplied to the purchaser at the time of delivery.

The statement of analysis must contain the following information expressed as percent by weight:

1. The total amount of nitrogen (N) in the fertilizer.
2. The amount of water insoluble nitrogen (WIN) present, if claimed.
3. The amount of available phosphate (P) present, expressed as P_2O_5.
4. The amount of soluble potassium (K) present expressed as K_2O.

Note that in the above example the total amount of nitrogen (N) is 12% of the net weight or six pounds. Half (6%), or three pounds of the total (12%) nitrogen present is in a water insoluble form. This means that half the nitrogen in the fertilizer is a slow release type.

If assistance is needed for understanding labels for applying fertilizer, consult your county agricultural educator of the UNH Cooperative Extension or the Extension turf specialist at the University of New Hampshire or the NH Department of Agriculture, Markets & Food, Division of Regulatory Services.

COMPLAINT RESOLUTION PROCESS

Even with good farm management, weather, schedules, and ground conditions may at times cause environmental and social problems from agricultural operations. In other cases, farm management and improper planning and implementation of practices may present longer term problems that need to be resolved. This section discusses the resolution of agricultural, environmental, and social problems.

In order to resolve environmental and social problems as quickly and efficiently as possible, and on a volunteer basis, RSA 431:33-35 provides a straight forward procedure to facilitate resolution. The RSA provides for a formal plan for handling complaints and actions to take if corrections are not made within 10 days after notification. If the person responsible fails to implement the recommended changes, the Commissioner of Agriculture, Markets & Food shall notify the Health Officer of the municipality and the Commissioner of Environmental Services for compliance action.

Complaints concerning manure, agricultural compost, chemical fertilizer handling, and other practices may be made by the public, adjacent landowners, municipalities, state agencies and others. The complaints may be made to towns, agencies, NHDES, etc., and are forwarded to the Commissioner of the NH Department of Agriculture, Markets & Food for resolution.

The Commissioner is empowered to investigate complaints of improper handling. This includes, but is not limited to, improper storage and spreading of manure, agricultural compost and chemical fertilizer. Water quality, air pollution, odor and nuisance may also be subjects of the complaint.

If the Commissioner is able to identify the source of the improper handling and has reason to believe such handling is a nuisance caused by failure to use BMPs, the Commissioner shall:

- Determine who is responsible for such handling
- Determine the changes needed in handling to comply with best management practices
- Notify, in writing, the person responsible for the problems and changes necessary to conform to best management practices
- Require a plan for compliance if the corrections, under RSA 431.35, I(c) have not been made within 10 days after notification

If the person responsible fails to implement the recommended changes, the Commissioner shall notify the Health Officer of the municipality and the Commissioner of the Department of Environmental Services who shall take such action as their authority permits.

Date of Complaint: ____________________ #________________
(Assigned by NHDAMF)

New Hampshire Department of Agriculture, Markets & Food

Best Management Practices
Complaint Form

Pursuant to RSA 431:33-35 Manure, Agricultural Compost, and Chemical Fertilizer Handling

Property Owner Information: (Person whom complaint is made against)

Name: ____

Address: ____

City/Town: ____________________State: ________Zip Code: ____________

Phone Number: ________________________

If known, times during the week when the property owner may be at the property:

____

Site Location of Alleged Mismanagement:

Address: ____

City/Town: ________________________

Directions to site from a major highway: Required: Please be specific:

____

Description of Alleged Mismanagement:

Describe the mismanagement; e.g. odors, water quality, etc. Include the number & type(s) of animals involved, names of nearby wetlands, streams or rivers, proximity to neighboring properties, runoff issues, type of material involved; manure, compost, chemical fertilizer. Attaching a map, photographs or a diagram is helpful.

____

Complainant Information (person making complaint):

Name: ____

Address: ____

City: ____________________ State: __________ Zip: ______________

Phone Number: ______________ Fax: ______________ Email: ______________

Print Name: ____________________ Signature(required): ____________________

Do you wish your name to be kept confidential by NHDAMF: [ ] Yes [ ] No **see note

For NHDAMF Office Use:

Date Complaint Received: ______________

Inspector: ____________________ On Site Inspection Date: ______________

Comments: ____

____

____

A complaint MUST be submitted in writing and signed by the complainant BEFORE a site visit can be done.

**We can not guarantee that complainant information can be kept confidential.

Forward this complaint form to: NH Dept. of Agriculture, Markets & Food, Div. of Regulatory Serv.
25 Capitol St., PO Box 2042
Concord NH 03302-2042
Phone: (603) 271-3685, 271-2753
Fax: (603) 271-1109
Email: registrations@agr.nh.gov

5/12/2011

TITLE XL
AGRICULTURE, HORTICULTURE AND ANIMAL HUSBANDRY

Chapter 431: Soil Conditioners

Manure, agricultural compost, and chemical fertilizer handling.

431:33 Definitions. Is this subdivision:

I. "Agricultural compost" means mixtures of decomposing organic matter, that may include manure, stored and managed on farms and used to condition and fertilize soils used for production of agricultural crops.

II. "Commercial fertilizer" means any substance containing one or more recognized plant nutrients which is used for its plant nutrient content and which is designed for use or claimed to have value in promoting plant growth, except unmanipulated animal and vegetable manures, marl, lime, limestone, wood ashes, and other products exempted by rule by the commissioner.

III. "Manure' means the excreta of animals, including poultry, that are kept or harbored as domesticated animals together with bedding materials used in housing such animals.

Source. 1991, 13:1, eff. June 18, 1991. 2005, 59:1, eff. July 22, 2005

431:34 Best Management Practices. In consultation with the agricultural advisory board, the commissioner of environmental services, the United States Natural Resources Conservation Service, the New Hampshire agricultural experiment station, the university of New Hampshire cooperative extension, and other appropriate agencies, the commissioner of agriculture, markets, and food shall identify and publish the best management practices for handling manure, agricultural compost, and commercial fertilizer. Such practices shall be based upon the best available research and scientific data so as to permit the maximum use of nutrient and soil conditioning values, while achieving the least possible adverse impact upon the environmental or human, animal and plant health.

Source. 1991, 13:1, 1995, 130:5; 134:23; 206:2, eff. Aug. 11, 1995. 2005, 59:2, eff. July 22, 2005.

431:35 Improper Manure Handling.

I. The commissioner shall investigate complaints of improper handling of manure, agricultural compost, and commercial fertilizer, including, but not limited to, complaints of improper storage and spreading. If the commissioner is able to identify the source of the improper handling and has reason to believe such handling is a nuisance caused by failure to use the best management practices, the commissioner shall:

a) Determine who is responsible for such handling.

b) Determine the changes needed in handling to comply with best management practices.

c) Notify, in writing, the person responsible of the findings and changes necessary to conform to best management practices.

d) Require a plan for compliance if the corrections, under RSA 43:35, I(c), have not been made within 10 days after notification.

II. If the person responsible fails to implement the recommended changes, the commissioner shall notify the health officer of the municipality and the commissioner of environmental services, who shall take such action as their authority permits.

Source. 1991, 13:1. 1996, 229:1. 1998, 60:2, eff. July 11, 1998. 2005, 59:3, eff. July 22, 2005.

NUISANCE LIABILITY OF AGRICULTURAL OPERATIONS

432:32 Agricultural Operation. "Agricultural Operation" when used in this subdivision includes any farm, agricultural, or farming activity as defined in RSA 21:34-a.

432:33 Immunity from Suit. No agricultural operation shall be found a public or private nuisance as a result of changed conditions in or around the locality of the agricultural operation, if such agricultural operation has been in operation for one year or more and if it was not a nuisance at the time it began operation. This section shall not apply when any aspect of the agricultural operation is determined to be injurious to public health or safety under RSA 147:1 or RSA 147:2.

432:34 Negligent or Improper Operations. The provisions of this subdivision shall not apply if a nuisance results from the negligent or improper operation of an agricultural operation. Agricultural operations shall not be found to be negligent or improper when they conform to federal, state, and local laws, rules, and regulations.

432:35 Limits. Nothing contained in this subdivision shall be construed to modify or limit the duties and authority conferred upon the Department of Environmental Services under RSA 485 or RSA 485-A or the Commissioner of Agriculture, Markets, and Food under any of the chapters in this title.

References

Bowman, J. S.1993. Dairy Fly Control Recommendations. University of New Hampshire, Durham, New Hampshire.

Maine, State of 1991. Strategy For Managing Nonpoint Source Pollution From Agricultural Sources and Best Management Systems Guidelines.

Mitchell, J. R. Selection and Use of Green Manure/Cover Crops. University of New Hampshire Cooperative Extension.

New Hampshire RSA 431:33-35. Manure, Agricultural Compost and Chemical Fertilizer Handling. Concord, New Hampshire.

Northeast Dairy Practices Council. Guidelines for Dairy Odor Control, NDPC46.

Northeast Dairy Practices Council. Guidelines for Handling Liquid Effluent from Milking Centers, NDPC15.

Northeast Regional Agricultural Engineering Service. On-Farm Composting Handbook, NRAES-54. Cooperative Extension.

Penn State. Streambank Fencing. Extension Circular 397.

University of New Hampshire Cooperative Extension. Instructions On How To Take A Soil Sample. University of New Hampshire, Durham, New Hampshire.

USDA-Natural Resources Conservation Service. Soil Survey (for each county in New Hampshire). Many are on-line at www.nh.nrcs.usda.gov, click on NH Soil Information.

USDA-Natural Resources Conservation Service, 1992. Agricultural Waste Management Field Handbook, National Engineering Handbook Series, Part 651.

USDA-Natural Resources Conservation Service. New Hampshire Technical Guide Handbook. Majority is on-line at www.nh.nrcs.usda.gov, click on Technical Resources.

USDA-Natural Resources Conservation Service. Engineering Field Handbook (Manual) For Conservation Practices, National Engineering Handbook Series, Part 650.

Welsch, P. J. Riparian Forest Buffers. USDA Forest Service.

SOME AGENGIES PROVIDING TECHNICAL ASSISTANCE

Technical assistance for agricultural Best Management Practices is available from the following agencies as resources permit:

New Hampshire Department of Agriculture, Markets, and Foods

Office of Commissioner
25 Capitol Street, PO Box 2042, Concord, New Hampshire 03302-2042
(603) 271-3551
Divisions – Refer to inside front cover.

University of New Hampshire Cooperative Extension

Office of the Director
Taylor Hall, 59 College Road, University of New Hampshire,
Durham, New Hampshire 03824-3587
(603) 862-1520
County Offices – Refer to inside front cover.

USDA-Natural Resources Conservation Service

State Office
Federal Building, 2 Madbury Road, Durham, New Hampshire 03824-2043
(603) 868-7581
NRCS Field offices, Conservation District offices, and Resource Conservation and Development Area Council offices – refer to inside back cover.

New Hampshire Department of Environmental Services

29 Hazen Drive, PO Box 95, Concord, New Hampshire 03302-0095
(603) 271-3503
Divisions and Bureaus – Consult with Department Operator at (603) 271-3503.

BMP References Available for Other New Hampshire Landuses

Stormwater Management and Erosion and Sediment Control Handbook for Urban and Developing Areas in New Hampshire. Rockingham County Conservation District, New Hampshire Department of Environmental Services, and Natural Resources Conservation Service. August 1992.

Best Management Practices for Erosion Control on Timber Harvesting Operations in New Hampshire – Resource Manual. J. B. Cullen, New Hampshire Department of Resources and Economic Development. (undated)

Best Management Practices for Controlling Soil Erosion on Timber Harvesting Operations in New Hampshire – A Pocket Field Guide for Foresters, Landowners, and Loggers. New Hampshire Timberland Owners Association. (undated)

Good Neighbor Guide for Horse-Keeping: Manure Management. University of New Hampshire Cooperative Extension, NH Department of Agriculture, Markets & Food, NH Department of Environmental Services, Natural Resources Conservation Service. April 1990.

Pesticide Management Guidelines For Groundwater Protection. University of New Hampshire Cooperative Extension, New Hampshire Department of Agriculture, Markets & Food, Division of Pesticide Control. November 1992.

Nitrogen and Groundwater. Nancy E. Adams and Ralph M. Winslow, Jr. UNH Cooperative Extension. (undated)

Keeping Barnyard Runoff in Its Place. NH Department of Environmental Services Greenworks. April 1997.

Timber Harvesting and Water Quality. NH Department of Environmental Services. Fact Sheet WD-WSEB 22-4, 1997.

Best Management Practices (BMPs) for Groundwater Protection. NH Department of Environmental Services. Fact Sheet WD-WSEB-22-4, 1997.

Planting Shoreland Areas. Ralph M. Winslow Jr., UNH Cooperative Extension. April 1995.

Composting Yard Waste. NH Department of Environmental Services. Fact Sheet WMD-SW-3, 1996.

Fertilization of Homelawns. John M. Roberts, UNH Cooperative Extension. 1996.

Best Management Practices to Control Nonpoint Source Pollution: A Guide for Citizens and Town Officials. NH Department of Environmental Services, NHDES-WD-97-8. Revised November 1997.

APPENDIX A:
Fencing Plans

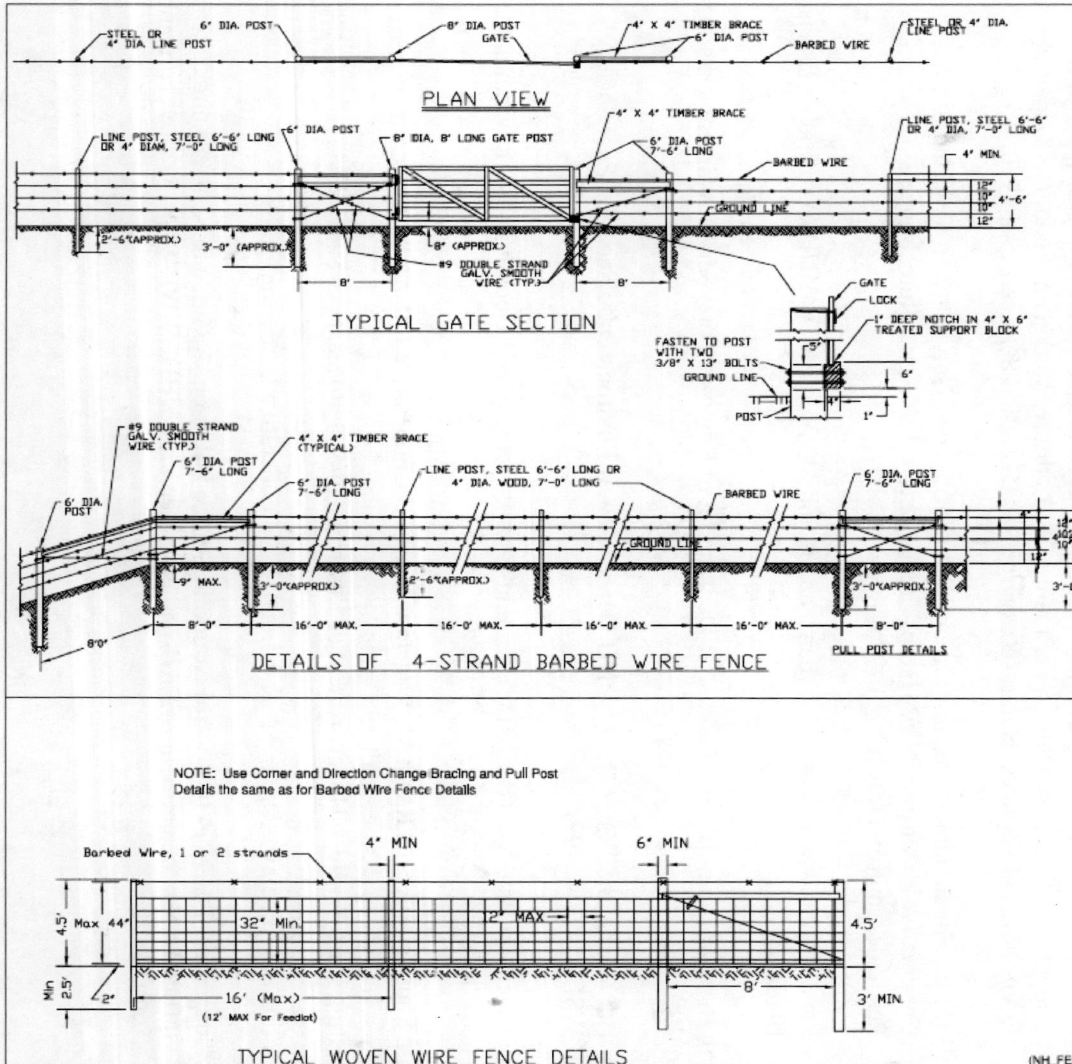

CONSTRUCTION DETAILS

POSTS:

1. All posts shall be pressure treated. Steel line posts shall be galvanized steel posts.
2. Always use wood posts for brace or corner posts.
3. Drive steel posts until flange is buried.
4. Place posts approximately: a) 8' apart for corner, gate, and pull post assemblies, b) 16' apart (maximum) when using woven or barbed wire.
5. Drive or bury posts at least: a) 2' for in-line fences, b) 3' (minimum) where bracing will be required, c) 3' (minimum) in corners.

BRACING:

1. Use pull post assembly for middle bracing approximately every 650' with in-line fences.
2. Use pull post assembly at gates, ends of fences, where the grade changes more than 15 degrees and the tension of the adjacent sections of wire is upward, and when alignment changes by 15 degrees or more.

FENCING:

1. Use galvanized woven wire, 14 gauge or heavier, 11 gauge or heavier for top and bottom wires, topped with at least one strand of barbed wire for control of hogs and sheep. Max. total height – 44".
2. Four strands of barbed wire 12½ gauge zinc coated or heavier with 14 gauge round 2-point barbes at approx 5" spacing may be used for controlling cattle or horses. Minimum height of 46".
3. Wire for bracing shall be No. 9 or heavier. Use 2 inch hardwood twist stock.
4. Staples shall be 9 gauge or heavier with a minimum length of 1½"; except for high tensile use 1¾" with slash cut points.
5. Fencing shall be placed on the side of the post opposite the area to be protected, except that fencing shall be placed on the outside of curves.
6. High tensile wire shall meet the following: 12½ gauge or heavier, type III galvanizing, with tensile strength 180,000 psi or greater. Use 6 strands minimum, and 8 strands for feedlots.
7. Always follow manufactures recommendations.
8. Provide one stop-post for each gate in the open position.
9. Gates shall be galvanized steel similar in design to those shown in "Typical Steel Gates" drawings.
10. Gates shall be hung with bottom hinge pin turned up and top hinge pin turned down.

Use Drawing Sheet "Fencing Details and Quantities" to tabulate Material needs and quantities.

FENCING DETAILS

NAME: ______

LOCATION: ______

New Hampshire Department of Agriculture
Ag. Nutrient Management Grant Program

Designed G. Lang — Date 4/05 — Approved by Gerald J. Lang, PE

Drawn — Traced — Checked — Title — Sheet No. of — Drawing No.

(NH_FENCE_DETAILS_2)

PERMANENT POWER FENCE DETAILS

ENERGIZERS

ELECTRONIC ENERGIZERS OR POWER FENCE CONTROLLERS SHALL BE INSTALLED ACCORDING TO THE MANUFACTURER'S RECOMMENDATIONS AND WILL MEET THE FOLLOWING MINIMUM REQUIREMENTS.

- HIGH POWER, LOW IMPEDANCE WITH A 5,000 VOLT PEAK OUTPUT AND A PULSE THAT IS LESS THAN 300 MILLIAMPS IN INTENSITY, FINISHED WITHIN 0.0003 OF A SECOND, AND AT A RATE OF 35-65 PULSES PER MINUTE.
- SOLID STATE CIRCUITRY.
- SAFETY PACE FUSE, TO PREVENT OVERPULSING.
- HIGH IMPACT, WEATHER RESISTANT CASE.
- JOULE RATING HIGH ENOUGH TO PROVIDE A MINIMUM SHOCK OF 2000 VOLTS FOR CATTLE, 3000 VOLTS FOR SHEEP AT THE POINT FARTHEST FROM THE ENERGIZER.
- BATTERY POWERED UNITS OF 3 JOULES OR MORE SHALL HAVE A SOLAR PANEL ATTACHED TO CHARGE THE BATTERY.

INTERIOR FENCES

MUST USE THE SAME WIRE AND POSTS AS THE PERIMETER FENCE. THEY MAY HAVE A SINGLE STRAND OF WIRE OR AS MANY 8 STRANDS OF WIRE, DEPENDENT UPON THE KIND AND CLASS OF LIVESTOCK.

LIGHTNING PROTECTION

TO REDUCE THE RISK DAMAGE, AN INDUCTION COIL MUST BE INSTALLED BETWEEN THE ENGERGIZER AND THE FENCE. THIS INDUCTION COIL IS OFTEN REFERRED TO AS A LIGHTNING CHOKE. ADDITIONAL LIGHTNING PROTECTION IS PROVIDED BY ATTACHING LIGHTNING DIVERTERS/ARRESTORS TO THE FENCE AND TO A 6 FT. LONG GALVANIZED ROD DRIVEN INTO THE GROUND. THIS GROUND ROD MUST BE LOCATED AT LEAST 65 FT. FROM THE GROUND RODS FOR THE ENERGIZER.

EARTH RETURN (GROUND RODS)

THE ENERGIZER GROUND WIRE MUST BE CONNECTED TO A GALVANIZED ROD 1/2 INCH OR LARGER IN DIAMETER, DRIVEN INTO THE GROUND A MINUMUM OF 6 FT. FOR EACH JOULE OF ENERGY OUTPUT. MULTIPLE EARTH RETURN RODS CONNECTED IN SERIES MAY BE USED, BUT THEY MUST BE SPACED AT LEAST 10 FT. FROM EACH OTHER. A MINIMUM OF 3 EARTH RETURN RODS ARE REQUIRED FOR ALL PERMANENT POWER FENCES. FOR BEST RESULTS EARTH RETURN RODS SHOULD BE INSTALLED IN AN AREA WHERE THE SOIL REMAINS MOIST. DO NOT USE COPPER GROUND RODS DUE TO CORROSION AT THE CONNECTION AND THE SUBSEQUENT LOSS OF ELECTRICAL CONTINUITY. TO PREVENT STRAY VOLTAGE, EARTH RETURN RODS MUST BE AT LEAST 65' FROM ELECTRIC UTILITY GROUND RODS, BURIED WATER PIPES, AND BURIED TELEPHONE LINES.

Table of Quantities

Material Item	Units	Qty
Posts, wood, 6" dia., 8' long	Ea.	
Posts, wood, 4" dia., 8' long	Ea.	
Posts, wood, 3" dia., 7' long	Ea.	
Posts, steel, "T" style, 6.5' long	Ea.	
Posts, fiberglass, 7/8" dia., 6' long	Ea.	
Wire, smooth, 12-1/2 gauge, zinc coated	Feet	
Galv. Steel Pins, 1/2" dia., 7" long	Ea.	
Insulators, wooden post mounted	Ea.	
Insulators, steel post mounted	Ea.	
Insulators, in-line assembly	Ea.	
Twist Stick, 1-1/2" – 2", 18" long	Ea.	
Wood, fiberglass, or polycarbonate stays	Ea.	

PLAN VIEW (NOT TO SCALE)

POST & WIRE SPACING FOR PERIMETER FENCES

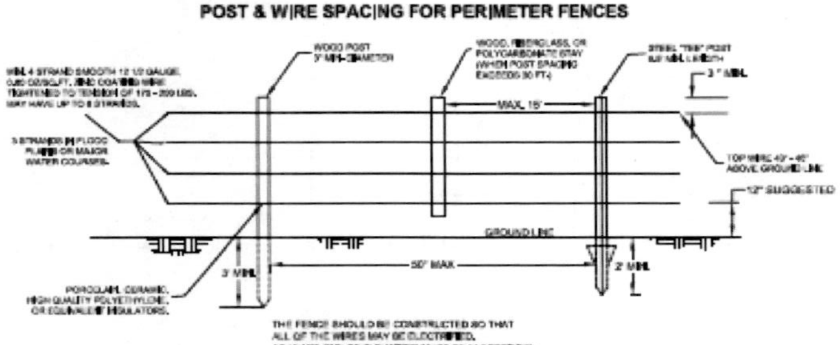

CORNER BRACE OR PULL ASSEMBLY

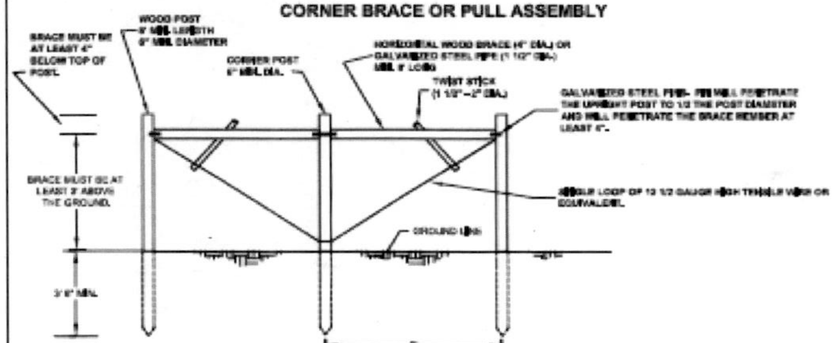

POSTS

WOOD POSTS

WOOD POSTS SHALL BE OF RED CEDAR, BLACK LOCUST, REDWOOD, PRESSURE TREATED PINE, OR OTHER WOOD OF EQUAL LIFE OR STRENGTH.
ALL WOOD POSTS EXCEPT RED CEDAR, BLACK LOCUST, REDWOOD, WHITE OAK, AND BURR OAK SHALL BE TREATED WITH CREOSOTE, PENTACHLOROPHENAL, OR COPPER CHROMATE ARSENATE (CCA) BY A METHOD SUCH THAT COMPLETE SATURATION OF THE SAPWOOD IS OBTAINED.
POSTS SHALL BE SOUND, NEW, AND FREE FROM DECAY WITH ALL LIMBS TRIMMED FLUSH WITH BODY.
USED RAILROAD TIES IN SOUND CONDITION MAY BE UTILIZED FOR POSTS IF FREE FROM EXCESSIVE CRACKING OR DECAY.

STEEL "TEE" POSTS

STEEL "TEE" WITH SUITABLE CORRUGATIONS, KNOBS, STUDS, OR GROOVES FOR FASTENING LINE WIRES CAN BE USED.
WIRE CLIPS WILL BE AS PROVIDED BY THE MANUFACTURER OR GALVANIZED WIRE, MIN. 12 GAUGE WILL BE USED.
EACH POST WILL HAVE AN ANCHOR PLATE ATTACHED.
THE POSTS WILL BE EITHER GALVANIZED OR PAINTED.
MIN. WT. : 133 LBS PER FOOT
MIN. LENGTH: 6.5 FEET
MIN. DEPTH INTO GROUND: 2 FEET

FIBERGLASS OR POLYCARBONATE POSTS

FIBERGLASS OR POLYCARBONATE POSTS CAN BE USED THEY SHALL BE AT LEAST 7/8" DIAMETER WITH ULTRAVIOLET PROTECTIVE COATING TO PROVIDE A 20 YEAR GUARANTEED LIFE AGAINST SPLINTERING OR DETERIORATING FROM SUNLIGHT OR WEATHER. THEY SHALL BE SET TO A MIN. DEPTH OF 18".

POSTS WILL BE DRIVEN INTO THE GROUND.

CORNERS ARE REQUIRED AT ALL POINTS WHERE THE FENCE ALIGNMENT CHANGES 15 DEGREES OR MORE.

PULL ASSEMBLIES ARE REQUIRED IN STRAIGHT SECTIONS OF THE FENCE SO THAT THE MAXIMUM DISTANCE BETWEEN BRACE POSTS DOES NOT EXCEED 660 FEET.

END BRACING IS REQUIRED WHERE THE FENCE ENDS, AND ON BOTH SIDES OF GATE OPENINGS.

ALL BRACE ASSEMBLY POSTS WILL BE DRIVEN INTO THE GROUND OR SET IN CONCRETE TO AVOID DISPLACEMENT WHEN TENSION IS APPLIED TO THE WIRES.

COOPERATOR APPROVAL:

I HAVE REVIEWED AND UNDERSTAND THE PLANS AND SPECIFICATIONS AND AGREE TO COMPLETE THE WORK ACCORDINGLY. FAILURE TO MEET THESE PLANS AND SPECIFICAITONS MAY JEOPERDIZE ANY COST SHARE APPLIED FOR. I UNDERSTAND THAT IT IS MY RESPONSIBILITY TO SECURE ALL NECESSARY PERMITS AND LICENSES, AND TO COMPLETE THE WORK IN ACCORDANCE WITH ALL LOCAL, STATE, AND FEDERAL LAWS. MODIFICATIONS OF THESE PLANS OR SPECIFICATIONS MUST BE APPORVED BY THE NATURAL RESOURCES CONSERVATION SERVICE BEFORE INSTALLATION. I ASSUME RESPONSIBILITY FOR NEGOTIATIONS AND AGREEMENTS WITH THE CONTRACTOR.

NAME DATE

BEFORE START OF CONSTRUCTION, THE OWNERS OF ANY UTILITIES INVOLVED MUST BE NOTIFIED. THE CONTRACTOR IS RESPONSIBLE FOR GIVING NOTICE BY CALLING "GOPHER STATE ONE-CALL" AT (800) 252-1166 AT LEAST 48 HOURS PRIOR TO ANY EXCAVATION.

MINNESOTA SPECIFICATIONS FOR CONSERVATION PRACTICES APPLY FOR ALL MATERIALS AND CONSTRUCTION WORK.

ALL MATERIAL USED SHALL BE NEW AND HAVE A MINIMUM LIFE EXPECTANCY OF 20 YEARS.

END ASSEMBLY

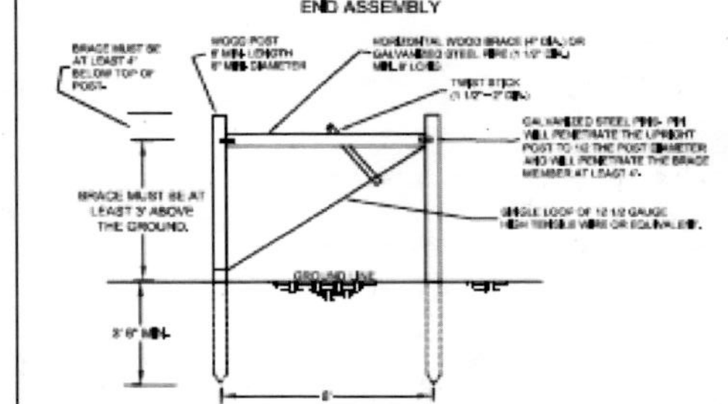

PERMANENT POWER FENCE DETAILS

Ag. Nutrient Management Grant Program

New Hampshire Department of Agriculture and Natural Resources Conservation Service

NAME:

LOCATION:

Gerald J. Lang, PE

SHEET OF

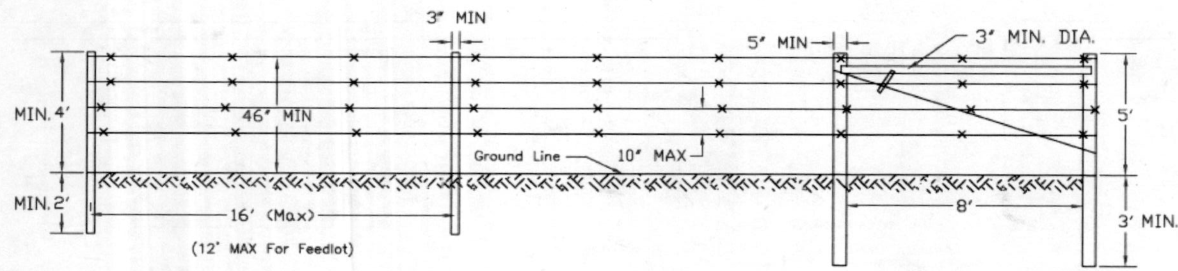

TYPICAL BARBED WIRE FENCE

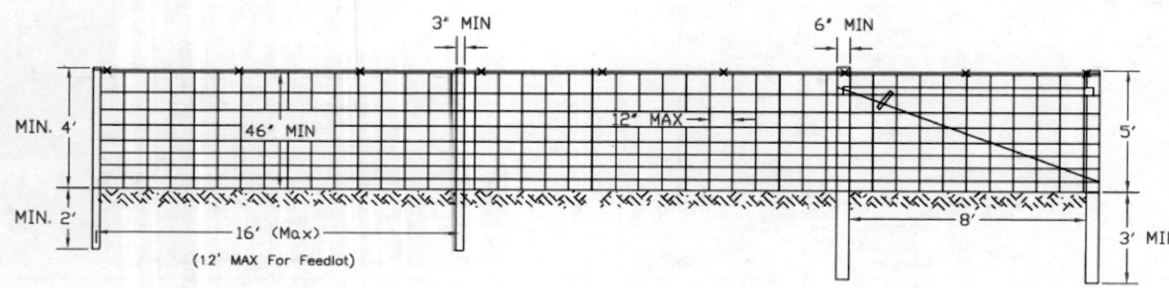

TYPICAL WOVEN WIRE FENCE

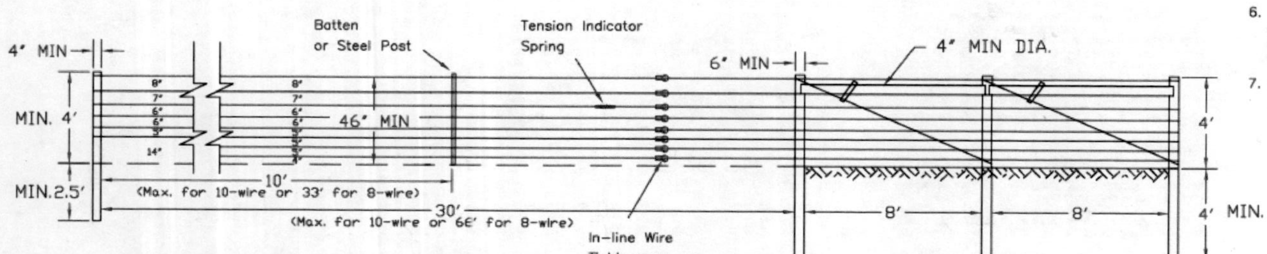

NOTE: Determine maximum spacing between primary posts using FOTG Standard 382

TYPICAL HIGH TENSILE FENCE

CONSTRUCTION NOTES

Posts

1. All posts shall be pressure treated softwood, black locust, eastern red cedar, or treated crossties.
2. Always use wood posts for brace or corner posts.
3. Drive steel posts until flange is buried.
4. Place posts approximately: a) 8 feet apart for corner, gate, and pull post assemblies, b) 16 feet apart (maximum) when using woven or barbed wire.
5. Drive or bury posts at least: a) 2' for in-line fence. b) 3' (minimum) where bracing will be required. c) 3' (minimum) in corners.

Bracing

1. Use pull post assembly for middle bracing approximately every 650 ft. with in-line fence.
2. Use pull post assembly at gates, ends of fence, where the grade changes more that 15 degrees and the tension of the adjacent sections of wire is upward, and when alignment changes by 15' or more.

Fencing

1. Use galvanized woven wire, 14 guage or heavier, 11 guage or heavier for top and bottom wires, topped with at least one strand of barbed wire for control of hogs and sheep. Min. height 44".
2. Four strands of barbed wire 12 1/2 guage or heavier may be used for controlling cattle or horses. Minimum height 46 inches.
3. Wire for bracing shall be No. 9 or heavier. Use 2 inch hardwood twist stock.
4. Staples shall be 9 gauge or heavier with a minimum lenght of 1 1/2"; except for high tensile use 1 3/4" with slash cut points.
5. Fencing shall be placed on the side of the post opposite the area to be protected, except that fencing shall be placed on the outside of curves.
6. High tensile wire shall meet the following, 12 1/2 gauge or heavier, type III galvanizing, with tensile strength 180,000 psi or greater. Use 6 strand minimum, and 8 strand for feedlots.
7. Always follow manufactures recommendations.

TYPICAL FENCING DETAILS

Name: ______

Location: ______

New Hampshire Department of Agriculture and Natural Resources Conservation Service

	DATE		DATE
DESIGNED		APPROVED BY G. Lang, PE	4/05
DRAWN		TITLE	
TRACED		CONST. APPROVAL	
CHECKED		TITLE	
		SHEET No. OF	

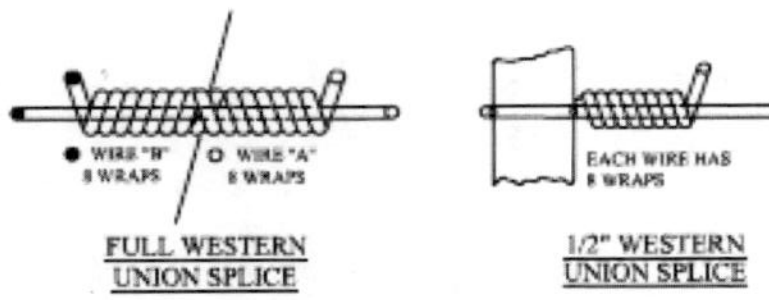

WIRE SPLICING DETAIL

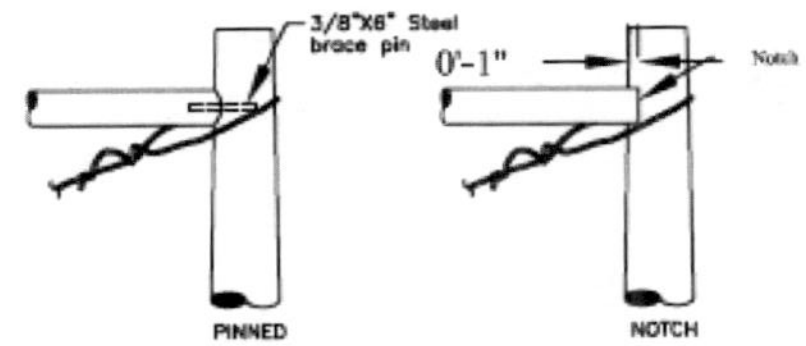

POST & BRACE DETAIL

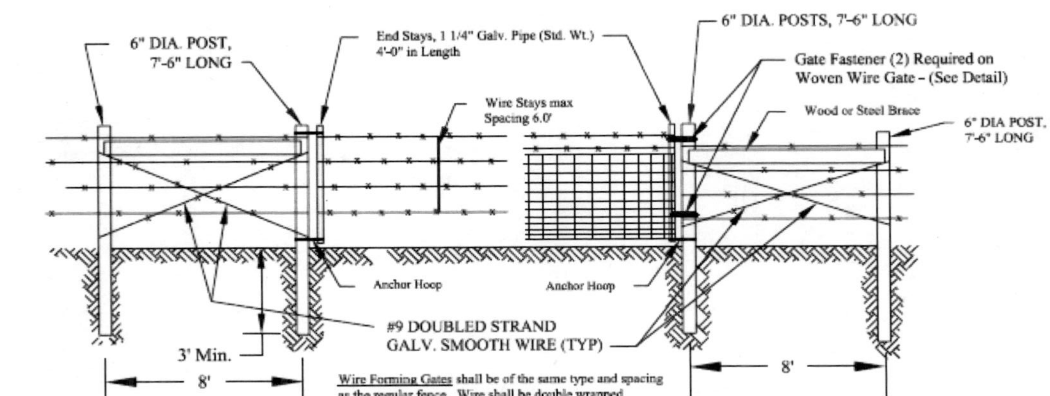

Anchor Hoops shall be two strands of 9 gauge wire stapled to the back side of the post.

Wire Forming Gates shall be of the same type and spacing as the regular fence. Wire shall be double wrapped around end stays and secured in place by drilling ¼" holes at bottom and top strands of woven wire and at each strand of barbed wire, tying these strands with 14 gauge galv. wire through the holes.

WIRE GATE DETAIL

Material and Quantity List

ITEM	UNITS	QTY
Post, End, Corner, or Brace; 6" dia., 7'-6" long	Ea.	
Post, Gate Hanging, 8" dia.,8'-0" long	Ea.	
Post, Line, wood, 4" dia., 7'-0" long	Ea.	
Post, Line, Steel, Galvanized, 6'-6" long, "T" Shape	Ea.	
Brace, wood, 4"x4" or 5" dia., 7'-6" long	Ea.	
Brace, steel, galvanized, 1 ½" dia., 7'-6" long	Ea.	
Dowel Pins, steel, ½" x 6" long	Ea.	
Wire, Barbed, Galv., 12 ½ gauge	Feet	
Wire, Woven, Galv., 14 and 11 gauge	Feet	
Wire, High Tensile, galv, 12 ½ gauge, 180,000 psi	Feet	
Wire, Tension, galv., 9 gauge	Feet	
Wire Stays, galv., 9 ½ gauge	Feet	
Staples, galv., 9 gauge, slash cut points	Lbs.	
Poles, wooden, 4" - 5" dia., 12' long	Ea.	
Planks, wooden, 2" x 6", 12' long	Ea.	
Spikes, galv., pole use, 6" long	Ea.	
Nails, galv., plank use, 3 ½" long	Lbs.	
Gates, pre-fab, max openming width = 16"	Ea.	
Twist Stick, 1 ½ - 2" dia., 18" long	Ea.	
Concrete (as needed), 3000 psi	cu. yds.	
End Stays, wire gate, 1 ¼ " dia., galv. pipe, 4' long	Ea.	

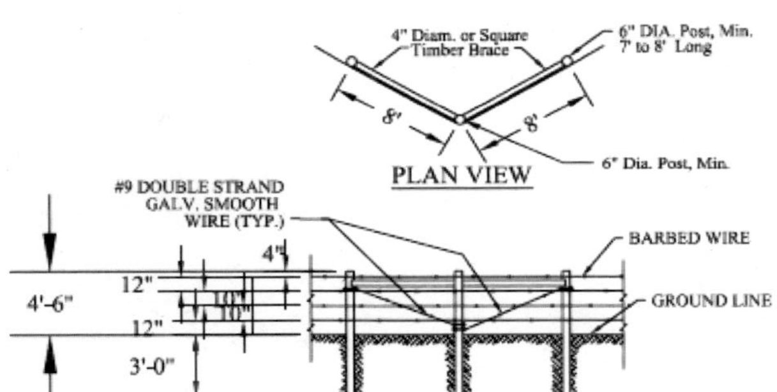

TYPICAL CORNER AND DIRECTION CHANGE BRACING

(NH_FENCE_Details&Qtys)

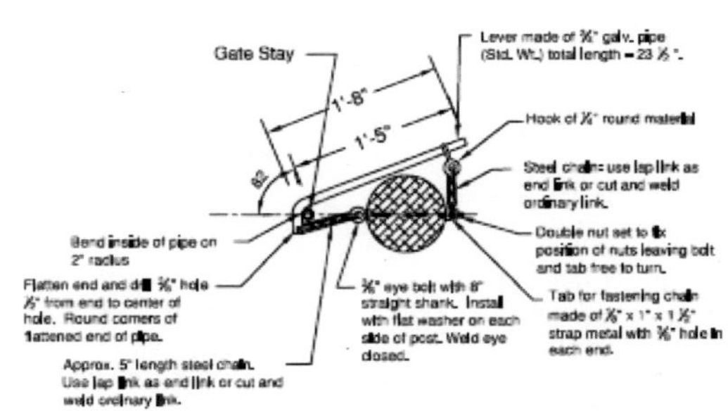

WIRE GATE FASTENER DETAIL

Fencing Details and Quantities

NAME: ______

LOCATION: ______

NH Department of Agriculture
Ag. Nutrient Management Grant Program

Designed G. Lang	Date 4/2005	Approved by Gerald J. Lang, PE
Drawn		Title
Traced		Sheet No. of / Drawing No.
Checked		

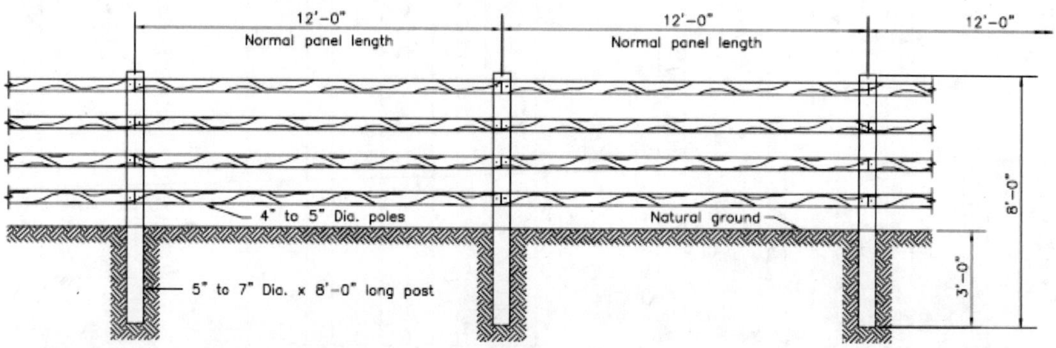

ELEVATION

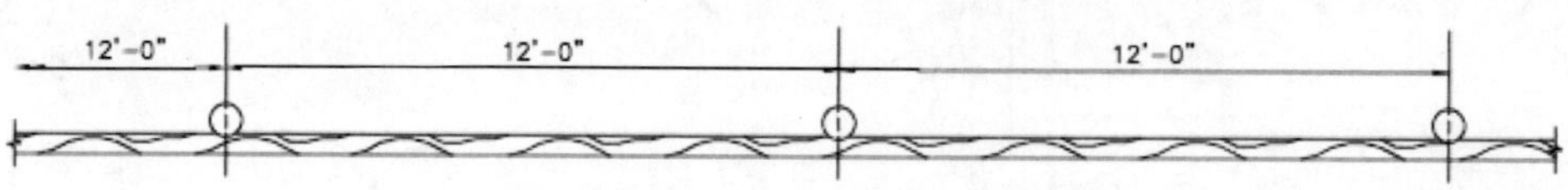

PLAN

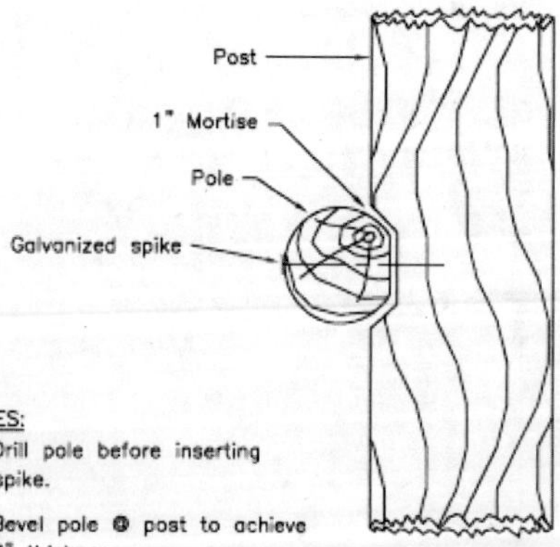

NOTES:

1. Drill pole before inserting spike.
2. Bevel pole @ post to achieve 3" thickness.

POST AND POLE CONNECTION

(NH_POLE_FENCE_DETAIL)

POLE FENCE

Name: ______

Location: ______

NEW HAMPSHIRE DEPARTMENT OF AGRICULTURE
Ag. Nutrient Management Grant Program

DESIGNED ______
REVIEWED ______
APPROVED Gerald J. Lang, PE

DRAWN	SCALE NONE
DATE	SHEET OF
DRAWING NO.	

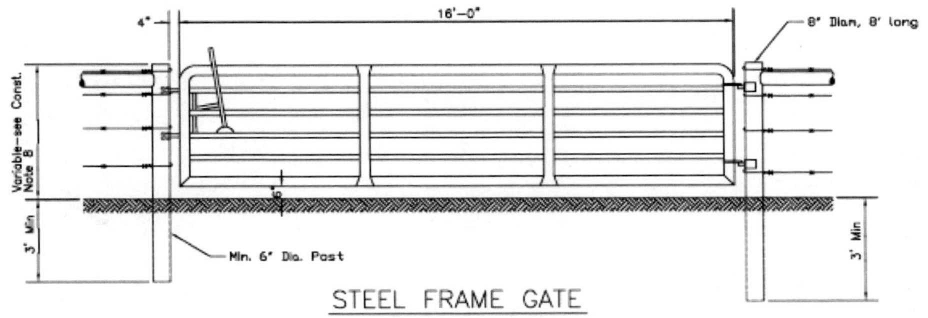

STEEL FRAME GATE

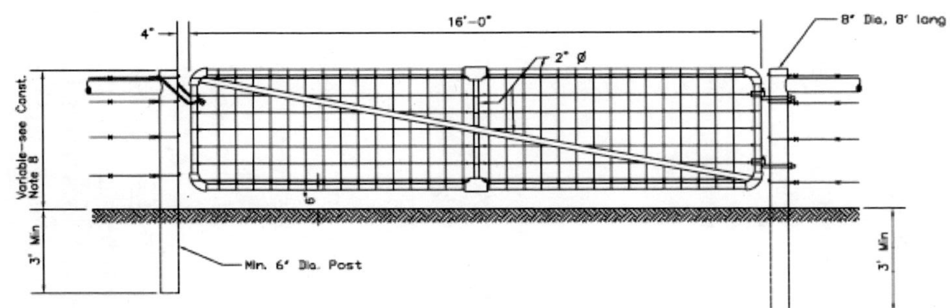

PIPE FRAME GATE

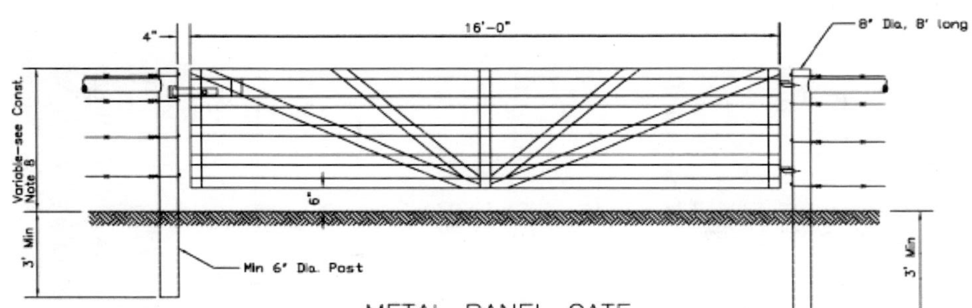

METAL PANEL GATE

MATERIALS NOTES

1. ALL MATERIALS SHALL BE SOUND, NEW, AND FREE OF DECAY AND RUST.
2. TREATED POSTS SHALL BE COMMERCIALLY AVAILABLE PRESSURE TREATED.
3. STEEL LINE POSTS SHALL BE 'TEE' TYPE WITH SUITABLE APPURTENCES FOR FASTENING LINE WIRES AND SHALL HAVE AN ATTACHED ANCHOR PLATE. STEEL POSTS WILL BE PAINTED OR GALVANIZED AND HAVE A MINIMUM WEIGHT OF 1.33 POUNDS PER FOOT EXCLUSIVE OF ANCHOR PLATE.
4. BARBED WIRE SHALL BE TWO TWISTED STRANDS OF 12 1/2 GAUGE OR HEAVIER WIRE WITH 14 GAUGE OR HEAVIER TWO POINT BARBS ON APPROXIMATELY 5 INCH CENTERS. ZINC COATING IS REQUIRED.
5. HINGES AND LOCKS SHALL BE INSTALLED AS SPECIFIED BY GATE MANUFACTURER.

CONSTRUCTION NOTES

1. CORNER, END, BRACE, AND PULL POSTS SHALL BE LOCATED AND SET FIRST.
2. STEEL POSTS SHALL BE USED AT LEAST EVERY 100 FEET FOR GROUNDING PURPOSES. STEEL POSTS MAY BE USED IN LINE CONSTRUCTION. LENGTH OF STEEL POSTS SHALL BE 6 FEET SET TO A DEPTH OF 2 FEET OR AT LEAST ONE INCH OVER THE ANCHOR PLATE, WHICH EVER IS GREATER.
3. LOCATION AND TYPE OF GATES ARE TO BE DETERMINED BY OWNER.
4. CUT NOTCH INTO GATE, BRACE, AND CORNER POSTS FOR HORIZONTAL BRACE. HORIZONTAL BRACE SHOULD BE PLACED APPROXIMATELY 3 FEET ABOVE GROUND.
5. STAPLE NO. 9 CROSS-BRACE, AND FENCE WIRES TO GATE, BRACE AND CORNER POSTS AT QUARTER POINTS OF POSTS.
6. WIRE STAYS ARE REQUIRED WHEN POST SPACING EXCEEDS 10 FEET. ONE STAY PER PANEL.
7. ADDITIONAL CONSTRUCTION REQUIREMENTS LOCATED IN CONSERVATION PRACTICE STANDARD 382 APPLY.
8. POST HEIGHT DIMENSION SHALL BE THE SAME AS REQUIRED FOR THE ADJACENT FENCE.

TYPICAL STEEL GATES

NAME: ______________________

LOCATION: ______________________

NEW HAMPSHIRE DEPARTMENT OF AGRICULTURE
Ag. Nutrient Management Grant Program

APPROVED BY Gerald J. Lang, PE

SOURCE: BLM DWG 02834-7

(NH_TYPICAL_GATES_DETAIL)

NO SCALE

APPENDIX B:
Manure Calculation Instructions
(Ag Nutrient Management Grant Program)

This Guide is intended for small animal farmers with 1 to 15 animal units. One animal unit is equivalent to 1 – 1000 pound animal. (Four 250 pound goats equals one animal unit). This guide is also intended for use on farms where the manure has considerable bedding and is fairly dry and stackable.

1. Column (1) and (2) - Use the form with animal type and weight closest to average weight of your animals. If the actual average weight is greatly different, contact your local NRCS office for more information.

2. Column (3) – Enter number of each type of animal next to appropriate weight class in Column (2).

3. Column (6) – Enter the number of days each animal type is confined between storage clean-out periods (typically 180 to 365 days). If animals are allowed outside part of the day, then multiply the confinement days by the decimal percent of time the animal is confined. (Example – Horse is outside in paddock 6 hours per day and storage is cleaned out every 6 months. Then confinement period = 180 days x (18/24) = 135 days).

4. Column (7) – Enter the product of column (3) x column (4) x column (6) to get the total quantity of manure produced in pounds between clean-out periods.

5. Column (8) – Enter the product of column (3) x column (5) x column (6) to get total volume of manure produced in cubic feet (Cu. Ft.) between each clean-out period.

6. Column (9) – Enter type of bedding used.

7. Column (10) – Enter the estimated quantity of bedding used per day per animal. Either weigh what you are using each day, take the amount used each week (pounds) and divide by 7 and then divide by the number of animals, or take the monthly amount used (pounds) and divide by 30 and then divide by the number of animals.

8. Column (11) – Enter the density value from Table at bottom of sheet.

9. Column (12) – Enter the product of column (3) x column (6) x column (10) all divided by column (11) to get the total volume of bedding used in cubic feet.

10. Column (13) – Enter the sum of column (12) divided by 2 plus (+) column (8) in cubic feet. We divide the bedding volume in half because the manure fills in the void spaces between the bedding (soaking up the liquid) and does not take up as much space as when it is dry. *Note: For horses, do not divide the bedding in half, if the stalls are cleaned-out more frequently than twice a month.*

11. Add up all the numbers in column (13) and put the total in the box at the bottom of the form. This is the total volume of manure that must be stored between clean-outs.

12. To compute the size of the storage facility, divide the computed Total Volume by the depth of the structure (4 feet maximum) – Total Volume ___________ / 4 = ___________ Square Feet. This is the bottom area needed for the facility. Divide this value by 8, 12, or 16 feet to get the length in feet – Bottom Area ___________ divided by 8, 12, or 16 = ___________ feet of length.

13. Now you can design the Storage Facility using the Standard Design Drawing with design tables for a "Post and Plank" type of structure with compacted mix of soil and rock dust or cement in the bottom. Concrete can be place in the bottom if the manure is more semi-solid with less bedding. The "Post and Plank" design is the preferred design, however, concrete "Waste Blocks" can be substituted for the "Post and Plank" design if desired and you have the equipment capable to handle them. *A 2' x 2' x 6' block weighs 3600 pounds – almost 2 tons and requires a large tractor and loader to move them.* The blocks can only be stacked two rows high and the bottom row must have soil backfilled against it to the middle of the block. Also alternate the placement of the second row so it covers half of two separate bottom blocks to help stabilize them.

14. If a roof is planned to cover the storage facility, then follow the Standard Design Drawing (2 sheets) for standard 12' long by variable widths with 12' being the maximum width before a center support post is needed. Standard Design is for a 12'x12' structure. Any wider structures will require center post(s) not to exceed a 12' span distance. Discuss with NRCS if a different design than the standard is needed.

Producer: ______________________

Animal Manure Production Bedding Used Calculation Worksheet

Date: ____________

Animal Type	Animal Weight (lbs)	No. of Animals	Manure Produced (lbs/day)	Manure Produced (cu ft/day)	No. of Days Confined per yr.	Total Manure in (lbs/yr)	Total Manure (cu ft/yr)	Bedding Type	Bedding Used (lbs/day/ animal)	Bedding Density (lbs/cu ft)	Total Bedding (Cu Ft)	Total Volume (Cu. Ft.)
(1)	(2)	(3)	(4)	(5)	(6)	(7)	(8)	(9)	(10)	(11)	(12)	(13)
					See Calc notes	(3)x(4)x(6)	(3)x(5)x(6)	See Calc notes	See Calc Notes	See Table below	[(3)x(6)x(10)] / (11)	(12) + (8) / 2
Horse	1000		45	0.7				(Shavings)	[illegible]	(9)		
Sheep	100		4	0.06				(Hay)	[illegible]	(4)		
Goats	100		4	0.06				(Hay)	[illegible]	(4)		
	150		6	0.09								
Chickens												
(Layers)	4		0.21	0.007				(Straw)	[illegible]	(2.5)		
(Broilers)	2		0.14	0.005								
Beef	500		30	0.46				(Straw)	[illegible]	(2.5)		
	750		45	0.7								
	1000		60	0.90								
Swine	65		4.2	0.06				(Straw)	[illegible]	(2.5)		
	150		9.8	0.15								
	200		13	0.2								
Dairy	1000		82	1.35				(Hay)	[illegible]	(4)		
	1400		115	1.8								
Other												
(Llama)	400		16	0.24				(Straw)	[illegible]	(2.5)		
(Alpaca)	400		16	0.24				(Straw)	[illegible]	(2.5)		
										Total Volume =		

Bedding Type:	Density:
Wood Shavings	9 lbs/Cu. Ft.
Straw	2.5 lbs/Cu. Ft.
Grass Hay	4.0 lbs/Cu. Ft.
Sawdust	12 lbs/Cu. Ft.

Typical Types and Amounts of Bedding used are shown in ( ) paranthesis.

NOTE: This sheet does not consider barnyard runoff or rainfall into the storage facility which could increase storage volume needed.

(See Calculation instructions on Sheet 2)

APPENDIX C:

Manure Storage/Compost Shed Plans

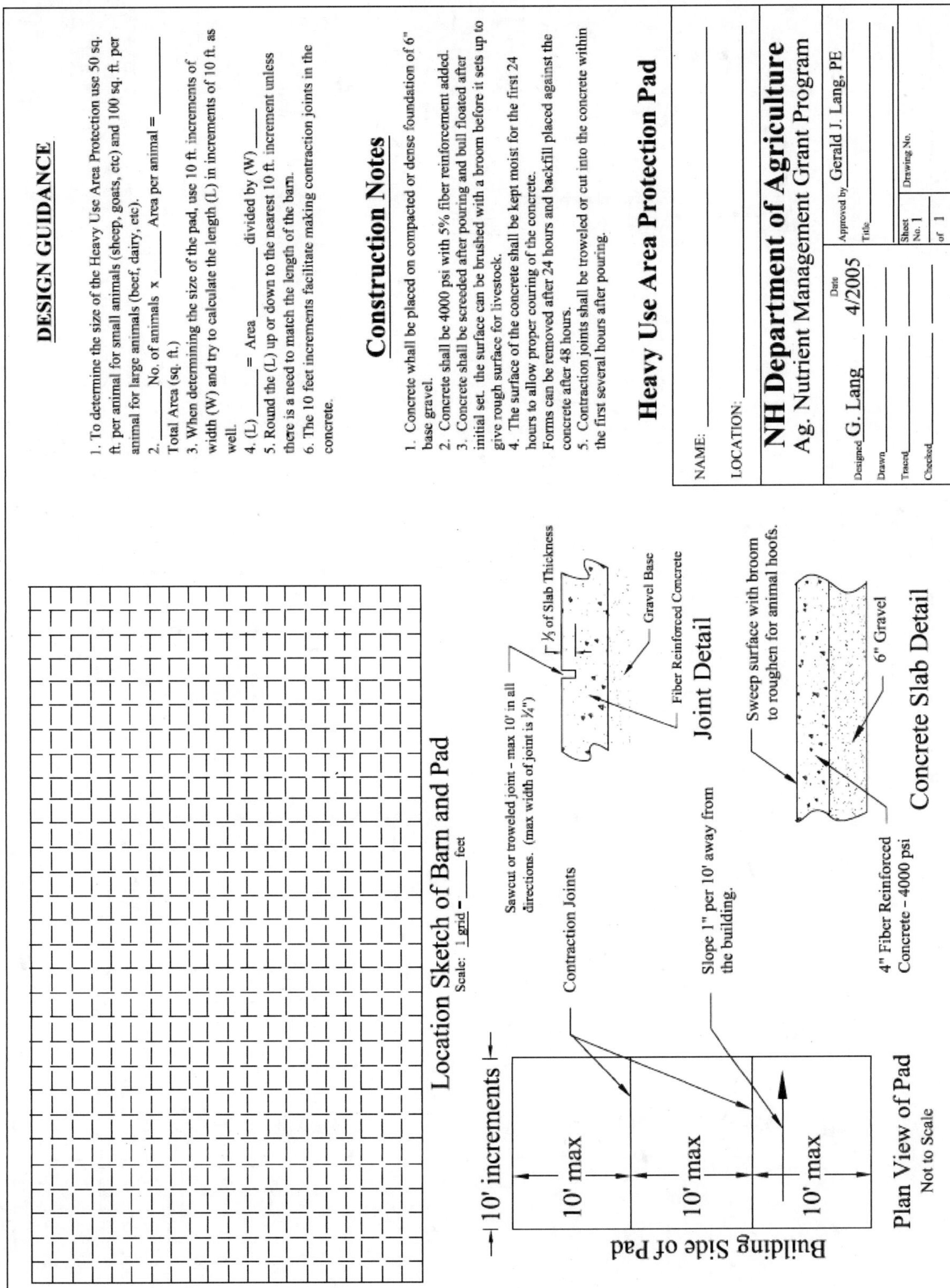

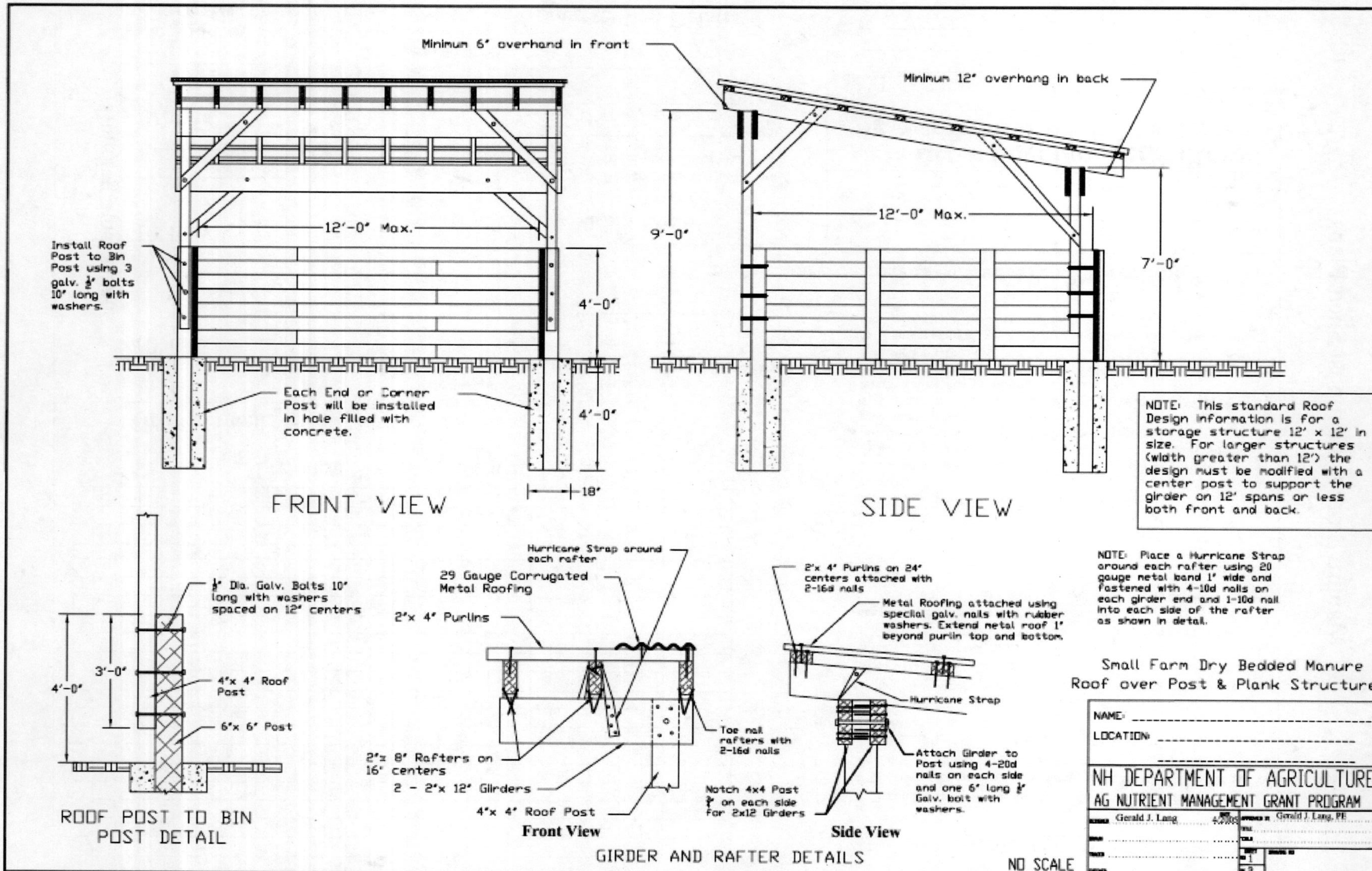
Minimum 6" overhang in front
Minimum 12" overhang in back
12'-0" Max.
12'-0" Max.
9'-0"
7'-0"
4'-0"
4'-0"
18"
Install Roof Post to Bin Post using 3 galv. bolts 10" long with washers.
Each End or Corner Post will be installed in hole filled with concrete.
FRONT VIEW
SIDE VIEW
NOTE: This standard Roof Design information is for a storage structure 12' x 12' in size. For larger structures (width greater than 12') the design must be modified with a center post to support the girder on 12' spans or less both front and back.
NOTE: Place a Hurricane Strap around each rafter using 20 gauge metal band 1" wide and fastened with 4-10d nails on each girder end and 1-10d nail into each side of the rafter as shown in detail.
Dia. Galv. Bolts 10" long with washers spaced on 12" centers
3'-0"
4'-0"
4"x 4" Roof Post
6"x 6" Post
ROOF POST TO BIN POST DETAIL
Hurricane Strap around each rafter
29 Gauge Corrugated Metal Roofing
2"x 4" Purlins
2"x 8" Rafters on 16" centers
2 - 2"x 12" Girders
4"x 4" Roof Post
Toe nail rafters with 2-16d nails
Front View
2"x 4" Purlins on 24" centers attached with 2-16d nails
Metal Roofing attached using special galv. nails with rubber washers. Extend metal roof 1" beyond purlin top and bottom.
Hurricane Strap
Attach Girder to Post using 4-20d nails on each side and one 6" long Galv. bolt with washers.
Notch 4x4 Post on each side for 2x12 Girders
Side View
GIRDER AND RAFTER DETAILS
Small Farm Dry Bedded Manure Roof over Post & Plank Structure
NAME:
LOCATION:
NH DEPARTMENT OF AGRICULTURE
AG NUTRIENT MANAGEMENT GRANT PROGRAM
Gerald J. Lang
Gerald J. Lang, PE
NO SCALE

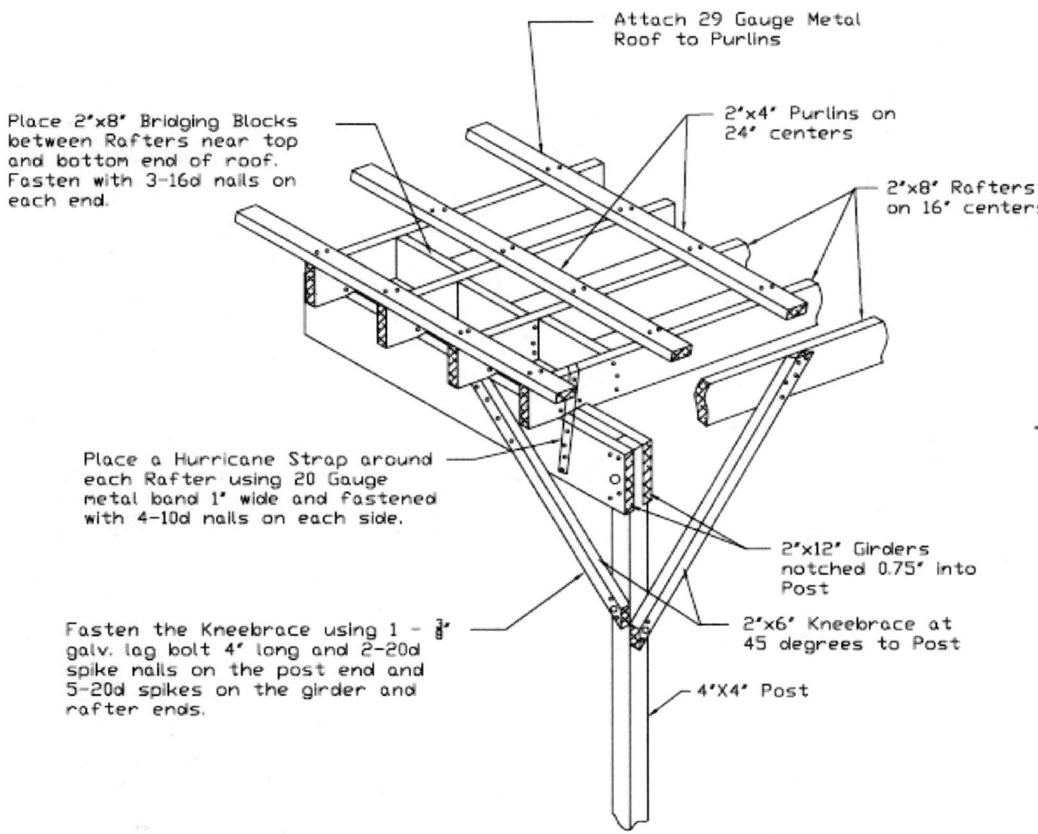

GIRDER TO POST AND RAFTERS DETAIL

MATERIAL LIST AND QUANTITIES

Item	Quantity	Unit
Planks 2"x 12" x 14' long		ea.
Planks 2"x 8" x 14' long		ea.
Planks 2"x 4" x 14' long		ea.
Posts 4"x4" x 8' long		ea.
20 gauge metal band 1" wide		Feet
Galvanized 20d Nails		lbs.
Galvanized 16d nails		lbs.
Galvanized ½" Bolts 10" long		ea.
Galvanized ½" Bolts 6" long		ea.
Galvanized ⅜" Lag Bolts 4" long		ea.
Galvanized 10d nails		lbs.
29 Gauge Metal roofing 14' long		Sheets
Concrete for Posts		Cu. Yds.
Planks - Kneebraces 2"x6" x 8'		ea.
Special Galv. Roofing nails		lbs.

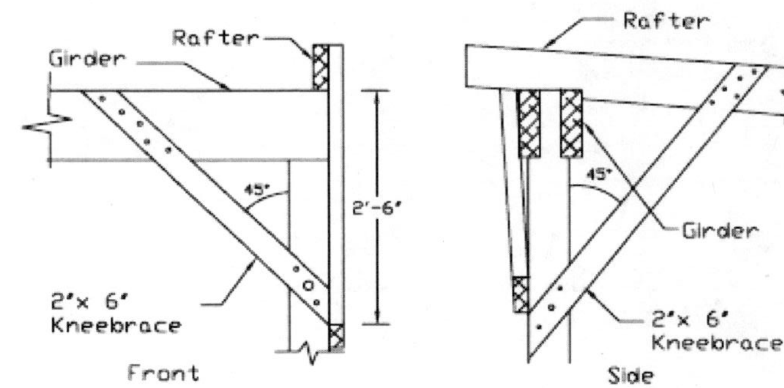

KNEEBRACE DETAILS

Small Farm Dry Bedded Manure
Roof over Post & Plank Structure

NAME: ______________________

LOCATION: ______________________

NH DEPARTMENT OF AGRICULTURE

AG NUTRIENT MANAGEMENT GRANT PROGRAM

Gerald J. Lang — Gerald J. Lang, PE

2 — 2

NO SCALE

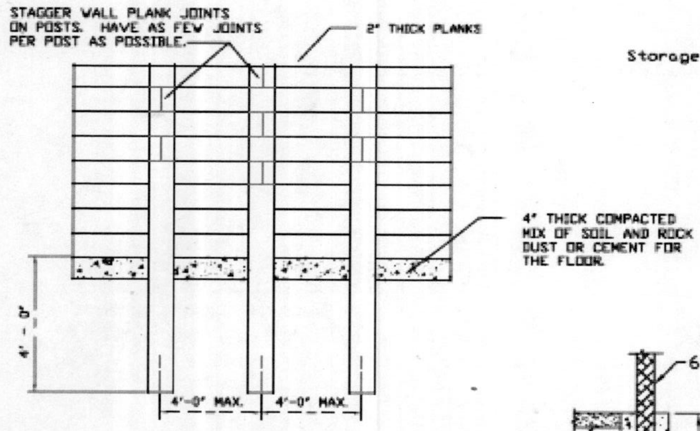

POST & PLANK WALL
WALL ELEVATION VIEW

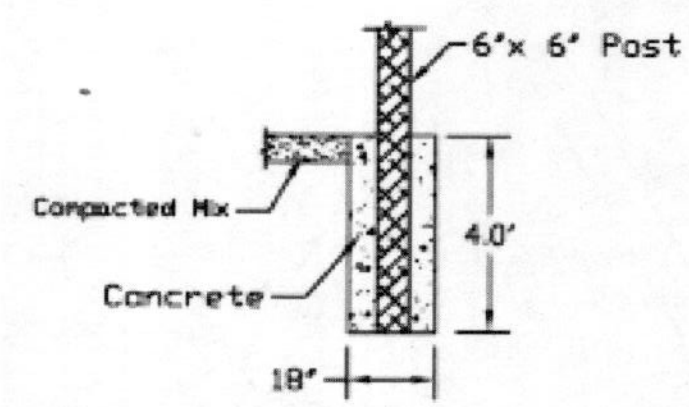

POST DETAIL FOR ROOF SUPPORT

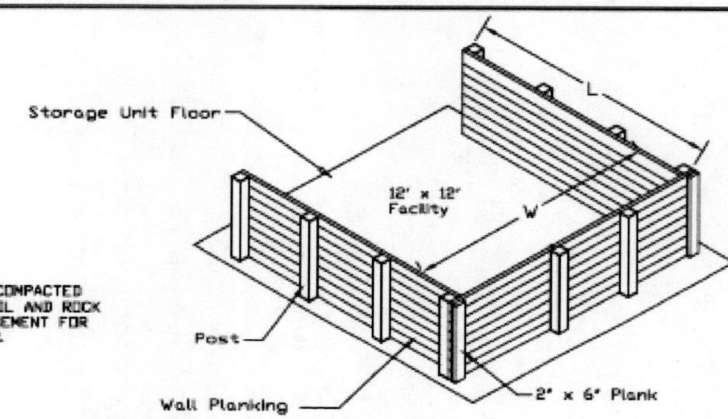

ISOMETRIC VIEW

TYPICAL 2-HORSE STORAGE SIZES

Days Between Emptying	Volume in (Cu. Ft.)	Floor Size in Ft. W x L
90	306	8' x 10'
180	612	12' x 12'
270	918	20' x 12'
365	1230	26' x 12'

Stone Dust or Cement		cu. ft.
Backfill material		cu. yds
Planks 2"x6" x 8' long		bd. ft.
Posts 6"x6" x 8' long		ea.
Bolts/Flat Washers		ea
Galvanized 20d Nails		lbs.
Planks 2"x6" x 4' long - corners		bd. ft.
Concrete (if needed) 4000 psi		cu. yds

NOTES:

1. All posts to be pressure treated with preservative.
2. All plank to be presssure treated with preservative.
3. Fasten each plank to each post with at least 2-20d galvanized spikes.
4. Post shall be no.1 Hemlock or equal
5. All nails (20d) and bolts and washers shall be galvanized.
6. The construction site shall be cleared of all trees, roots, sod, brush, rocks, and debris. All material not suitable for subgrade shall be removed and replaced with compacted earthfill at or above the planned final level.
7. The area 1.0' beyond the floor size will be re-worked with a roto-tiller after spreading a minimum of 1" of ROCK DUST or POWERED PROTLAND CEMENT over the surface.
8. After mixing, the area will be compacted with a weighted yard roller or fabicated tamper and smoothed level. Be sure the original soil surface is damp before beginning the re-work phase.
9. Prepare the floor area of the storage facility before laying out and digging post holes.
10. Posts shall be placed in post holes, aligned and braced, prior to tamping soil tightly around the posts.
11. Joints of wall planking shall be staggered between posts.
12. Earth backfill against the outside of the plank wall shall not be more than 12" above the floor level.
13. If a roof will be constructed over the Storage Facility, then the four end por corner posts must be installed using concrete around posts - see detail.
14. Widths greater than 12' will require mid posts to keep spans less then 12' both fron and back. Lengths shall not exceed 12'.

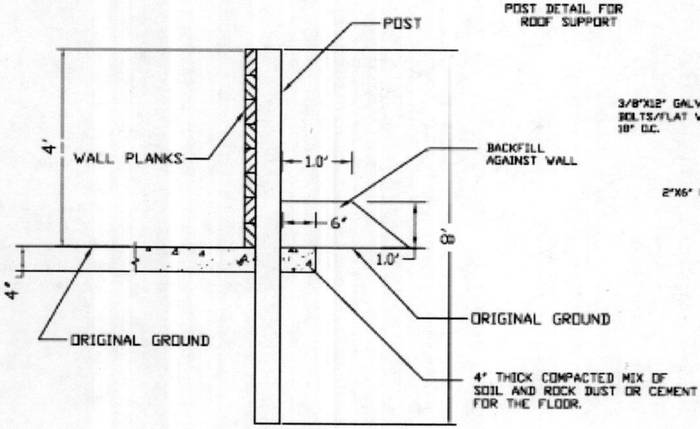

POST DETAIL

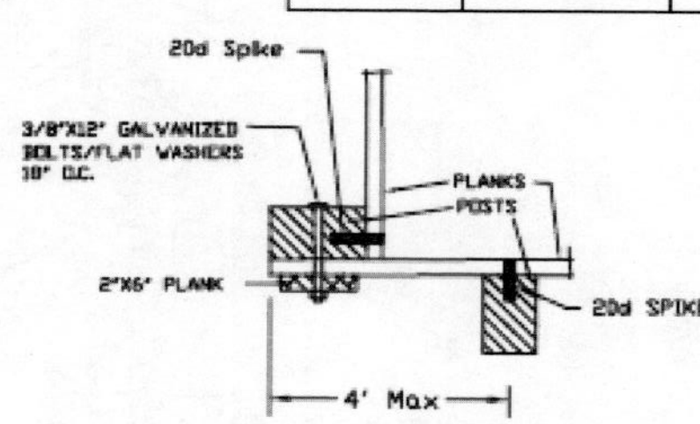

LEFT CORNER DETAIL

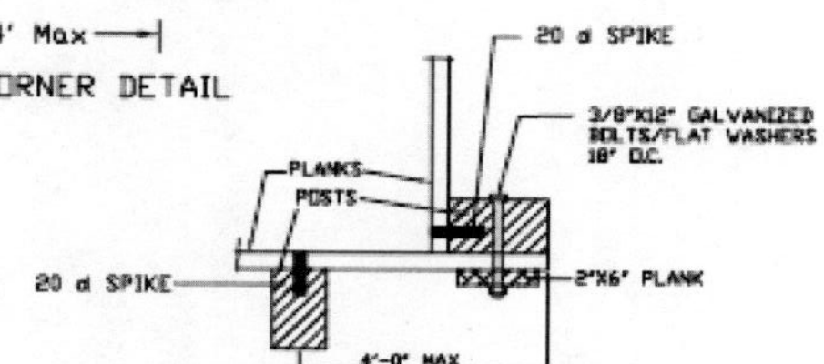

RIGHT CORNER DETAIL

Small Farm Dry Bedded Manure
Post and Plank Structure

NAME: ______________________
LOCATION: ______________________
____

NH DEPARTMENT OF AGRICULTURE
AG NUTRIENT MANAGEMENT GRANT PROGRAM

Gerald J. Lang 4/2005 Gerald J. Lang, PE

NO SCALE

APPENDIX D:
Composting Design Guide for Small Farms

Why Compost?

Gardeners and organic growers know the value of composted manure for improving soil fertility, tilth, and moisture holding capacity. Horse stall compost has almost a perfect proportion of nutrients to organic matter for garden crops. Horse and small animal owners should consider composting for other reasons as well. (1) Composting reduces the volume of waste to be stored on the farm. (2) Composting is a clean alternative to storing manure and bedding in piles. (3) The high temperatures created during composting kills fly eggs, and well-composted cleanings will not attract adult flies.

Composting Basics

Composting is easy. However, there are a few things to remember to produce quality compost. The recipe to produce good compost from horse manure and wood shavings only requires adding water and mixing. If the manure you are composting comes from other animals (beef, sheep, goats, etc) with less bedding, then you may need to add additional *carbonaceous materials* such as: sawdust, wood shavings, grass clippings, leaves, etc to improve the composting process.

Composting typically takes place in three phases. It takes 6 to 8 months to produce cured compost from raw stall waste. The phases are:

a) A short warm-up phase lasting a few days to a month.
b) A hot composting phase lasting 3 to 4 months.
c) A cool curing phase lasting another 3 to 4 months.

The following information provides additional suggestions for producing good compost:

1. Select a site for your composter bins that are easily accessible, level, and on dry ground. The bins should be at least 100 to 150 feet away form wells, ditches, streams, and lakes. Leave a buffer strip of taller grasses, wildflowers, and shrubs between the compost bins and any drainage way to filter stormwater runoff from the compost area. The compost piles should be covered to control and maintain proper moisture levels. A properly sized manure storage facility with roof could be used to compost manure. (Ex. A 12' x 12' bin under a roof with a center wall splitting the bin in half may be perfect for 6 months storage in each half allowing composting to take place in one half while storing manure in the other half.) A pile of the mixture on the ground with a tarp covering it can also work. Either way, management of the compost mixture with periodic turning is important.

2. Thoroughly mix the raw material with compost or several cubic feet of soil to begin the composting process.

3. Maintain 50% moisture content in the material by adding water as needed with a garden hose. (Squeezing moist compost should feel moist but not saturated with liquid.)

4. Start the process with enough material in the stack to insulate the hot compost. The pile should be at least 6 feet wide and 3 feet high.

5. If the pile is too wide or too high, the flow of oxygen to the micro-organisms composting the mixture may be cut off. Make sure the bin is less than 12 feet wide and 4 feet high.

6. Manure without bedding, or manure with sawdust or wood shavings may create a pile with too little air space. Measure air space using the "five-gallon bucket test" (see below). Add bulking materials, such as shredded wood, bark, or dry straw, to increase air space as needed.

7. Arrange and size the bins so they can be accessed easily for mixing and emptying. You will need a minimum of two bins to store and compost at the same time. Size each bin to store 6 to 8 months of produced waste. Bins can also be covered with tarps rather than a roof to control moisture.

8. Turning the mixture, mixes the pile's cooler outside layer with the hotter center and enhances the composting. After the pile is built, wait **7 to 28** days before turning it the first time so it can "cook". Try turning the pile again at **24, 72 and 120** days. **Three to seven** turns during the life of the pile are common. Base the turning schedule on the compost pile's materials, weather, and anticipated use of the compost. When mixtures have the right amount of moisture and air space, a temperature of 120° F or lower usually indicates the need to turn the pile so it can reheat.

9. The higher the bedding-to-manure ratio, the more likely it is that you will need supplemental nitrogen. If you have the proper amount of water and free air space and the mixture doesn't heat up, add one-third cup of a commercial nitrogen fertilizer to the pile each day until it begins to "cook".

10. Additional information for operating a composting facility is available from the Cooperative Extension or your local NRCS office. There is also a lot of information available on the World Wide Web.

Using Compost

Good quality compost should be applied only at recommended rates and to plants and soils that can use the nutrients. As a rule of thumb, good quality horse manure compost can be applied ½ to 1 inch thick and then mixed well into the soil. You will need to think about who will use the finished compost. Will you use it yourself, sell or give it to our neighbors, or market it in the local area. By using compost to grow plants we complete the organic matter cycle.

Five-gallon Bucket Test

Materials needed:

- Five-gallon pail
- One-gallon pail
- Typical mix of materials added to the compost mixture (horse manure, shavings, straw, etc)

1. Fill the five-gallon pail one-third full with the mixture of typical compost material. Drop the pail 10 times from a height of 6 inches onto a concrete floor or sidewalk. Be careful not to spill any of the compost material.

2. Add more material to fill the five-gallon pail two-thirds full. Drop the pail 10 times from a height of 6 inches as before.

3. Fill the five-gallon pail up to the top. Drop the pail 10 times from a height of 6 inches again. Fill the five-gallon pail to the top once again.

4. Add water to the five-gallon pail, keeping track of how much you can fit in before it overflows. If you can add 2-1/2 to 3 gallons of water, you have adequate free air space. If not, you need to add more bulking material such as straw, coarse wood chips, or shredded bark.

5. If you can add more than 3 gallons of water, you have too much free air space. The particle size must be reduced by shredding or grinding the compost materials or by mixing finer materials and/or manure into the compost.

Your Composting Operation

Composting is a balancing act. Providing ideal environmental conditions for microbial growth accelerates the process. Just enough water, air, carbon, and nitrogen getting piled, turned, and aged without contaminants makes for good compost.

You'll need to customize the process to fit your specific combination of manure, bedding, and other organic materials. You can find the best mixture by developing a clear understanding of the process, accurately measuring materials, and going through some trial and error. Mixing the compost pile is important to get good quality compost. Depending on the size of the pile, a tractor mounted front end loader works well to mix the pile in the bin. You may also need multiple bins depending on the amount of waste material produced to keep the maximum size less than 12'x12'. Several smaller widths are better than one large size.

United States Department of Agriculture
NRCS Natural Resources Conservation Service

NRCS
State Office
Federal Building, 2 Madbury Road
Durham, NH 03824-2043
(603) 868-7581
George W. Cleek, State Conservationist

BELKNAP COUNTY
Belknap County Conservation District
719 Main Street, Room 203
Laconia, NH 03246-2772
(603) 527-5880

CARROLL COUNTY
NRCS and Carroll County Conservation District
The Grindle Center, 73 Main St, PO Box 533
Conway, NH 03818-0533
(603) 447-2771

CHESHIRE COUNTY
NRCS and Cheshire County Conservation District
11 Industrial Park Drive
Walpole, NH 03608-9744
(603) 756-2988

COOS COUNTY
NRCS and Coos County Conservation District
4 Mayberry Lane
Lancaster, NH 03584-3616
(603) 788-4651

GRAFTON COUNTY
NRCS and Grafton County Conservation District
250 Swiftwater Road, Room 6
Woodsville, NH 03785-1424
(603) 747-2001

HILLSBOROUGH COUNTY
NRCS and Hillsborough County Conservation District
Chappell Professional Center
#468, Route 13, South
Milford, NH 03055-3476
(603) 673-2409 ext. #4

MERRIMACK COUNTY
NRCS and Merrimack County Conservation District
The Concord Center
10 Ferry Street, Box 312, Suite 211
Concord, NH 03301-5081
(603) 223-6023

ROCKINGHAM COUNTY
Rockingham County Conservation District
110 North Road
Brentwood, NH 03833-6614
(603) 679-2790

STRAFFORD COUNTY
Strafford County Conservation District
USDA Agricultural Service Center
259 County Farm Road, Unit #3
Dover, NH 03820-6015
(603) 749-3037

NRCS (ROCKINGHAM & STRAFFORD COUNTIES)
243 Calef Highway
Telly's Plaza
Epping, NH 03042-2326
(603) 679-1587

SULLIVAN COUNTY
Sullivan County Conservation District
24 Main Street
Newport, NH 03773-1515
(603) 863-4297

NH RESOURCE CONSERVATION AND DEVELOPMENT (RC&D) AREA COUNCILS

North Country RC&D Area Council
719 Main Street, Room 220
Laconia, NH 03246-2772
(603) 527-2093

Southern New Hampshire RC&D Area Council
10 Ferry Street, Box 4, Room 422
Concord, NH 03301-5019
(603) 223-0083